NanoScience and Technology

The series NanoScience and Technology is focused on the fascinating nano-world, mesoscopic physics, analysis with atomic resolution, nano and quantum-effect devices, nanomechanics and atomic-scale processes. All the basic aspects and technology-oriented developments in this emerging discipline are covered by comprehensive and timely books. The series constitutes a survey of the relevant special topics, which are presented by leading experts in the field. These books will appeal to researchers, engineers, and advanced students.

More information about this series at http://www.springer.com/series/3705

Mircea Dragoman · Daniela Dragoman

2D Nanoelectronics

Physics and Devices of Atomically Thin Materials

Springer

Mircea Dragoman
National Research and Development
Institute in Microtechnologies
Bucharest
Romania

Daniela Dragoman
Faculty of Physics
University of Bucharest
Bucharest
Romania

ISSN 1434-4904 ISSN 2197-7127 (electronic)
NanoScience and Technology
ISBN 978-3-319-48435-8 ISBN 978-3-319-48437-2 (eBook)
DOI 10.1007/978-3-319-48437-2

Library of Congress Control Number: 2016955307

Printed on acid-free paper

This Springer imprint is published by Springer Nature
The registered company is Springer International Publishing AG
The registered company address is: Gewerbestrasse 11, 6330 Cham, Switzerland

Preface

We have started to work in the area of graphene at the end of 2006, discovering that the fascinating Dirac equations could drive to new discoveries in solid-state physics. At that time, although the Dirac equation, which governs the transport of carriers in graphene, was discovered by Wallace long time ago (P.R. Wallace, The Band Theory of Graphite, Phys. Rev. 71, 622, 1947), many theoretical issues were still unsolved, and some of them are not solved even today. For instance, the problem of metal–graphene contact resistance, which should clarify how to match two physical systems in which carrier transport is described either by the Dirac equation (graphene) or by the Schrödinger equation (metal), is currently under investigation. In 2006, however, the experiments were in infancy, and the Manchester group led by A. Geim and K.S. Novoselov just published their papers about the amazing electrical properties of graphene (K.S. Novoselov et al., Electric Field Effect in Atomically Thin Carbon Films, Science 306, 666, 2004). In a subsequent major publication (M.I. Katsnelson, K.S. Novoselov, A.K. Geim, Chiral Tunneling and Klein paradox in graphene, Nature Physics 2, 620, 2006), the Manchester group has cited one of our finest scientific work (D. Dragoman, M. Dragoman, Optical Analogue Structures to Mesoscopic Devices, Progr. Quantum Electronics 23, 131, 1999) completely ignored at that time by the scientific community. This paper is at the foundation of our book Quantum-Classical Analogies, published in 2004 at Springer, and which can be found in hundreds of libraries worldwide according to WorldCat. The Manchester group has cited us because they discovered that the ballistic transport properties of graphene at room temperature, which manifest over large distances (from few hundreds of microns up to few microns depending on how graphene is encapsulated), can be applied to various electronic devices mimicking optics. Today, after ten years, we can say that we have successfully followed the path of ballistic electron transport-optics analogies, and in the last two years we have experimentally evidenced the first ballistic transistors working at room temperature with impressive electrical and optical properties, and we have developed theoretically the first quantum gates based on ballistic transport. Meanwhile, we have written tens of papers about graphene devices, reviews, and we have investigated together with our colleagues the first microwave devices based on graphene and the first photodetectors.

Many expectations regarding graphene were immediately spread all over the world, and many groups have started to work in this new area of science. It was thought that in few years graphene will replace silicon, and a new era in applied sciences, especially in nanoelectronics and photonics, will emerge.

However, at that time "the wonder material" could be barely found even in the form of flakes with maximum dimensions of 1 mm or graphene inks. We have obtained some graphene flakes from Peter Blake due to A. Geim, and in this way the long experimental work with graphene has started. We have understood from the beginning that the road to graphene devices will be long and hard. Despite very optimistic reports spread on the net and in many influential journals, we have known since 2008 that from the graphene flakes, which allows you to fabricate 2–3 devices, up to the high degree of integration of Si technology, which is able today to integrate more than 5 billions of transistors, is a long, very long way. Indeed, today integrated circuits based on graphene and containing 3–4 devices are a rarity, although graphene wafers of 4 in. and even 6 in. are fabricated by companies such as Graphene Laboratories, known also as Graphene Supermarket, or Graphene Industries, with which we cooperate since five years. Why? There are two main ways to grow graphene: via CVD, which requires a transfer of graphene grown on copper on a certain substrate, or via epitaxial growth on SiC at very high temperatures. If you examine such a wafer on SEM, you will find cracks, wrinkles, and defects in many places of the wafer. If we fabricate a device on such a wafer, the yield is 30–60% depending on the substrate. This is the main reason why graphene and many other atomically thin materials cannot compete with Si technology yet, although a lot of money is invested in the research. Moreover, if now graphene devices have gained a certain maturity, this is not the case for other 2D materials, where flakes are still used and devices at the wafer level are still a rarity.

We have reflected a lot if it is necessary to write a book about graphene and other 2D materials, their physics, and applications, since the subject is still immature despite thousands of published papers, and theoretical works prevail too much over experimental results. The best book about graphene theory in our opinion is the book of M.I. Katsnelson, Graphene-Carbon in Two Dimensions, Cambridge Univ. Press, 2012. However, a book about devices based on graphene and other 2D materials and their physics is still missing, the results being spread in papers and/or books edited by various editors.

The present book is divided into three chapters. The first and main chapter, containing about 50% of the book, is dedicated to graphene devices such as transistors, diodes, sensors, integrated circuits, optoelectronic circuits, and biosensors based on graphene. The remaining two chapters are dedicated to other 2D materials, especially to transition metal dichalcogenides (TMDs), and describe their physics, growth techniques, and devices based on them, such as transistors, photodiodes, tunneling diodes, memristors, and even elementary circuits. Many figures, maybe in excess, accompany the explanations. The book was written having in mind many

Ph.D. students, postdocs, and young researchers, which need to absorb a lot of new knowledge, not taught at universities, in a rather short time.

The book is written with the hope that this ultimate research area in applied physics, dealing with atomically thin materials, will evolve and develop in the next years. The maturity of the growth processes of graphene and other 2D materials is the main challenge for the years to come. No serious application will be developed until hundreds of devices will be fabricated on the same wafer, and the reproducibility will be comparable to that in semiconductors.

Graphene is zero-bandgap semiconductor. Therefore, designing and fabricating devices that are analogous to semiconductor devices is a straightforward, but many times unsuccessful way to develop new applications. The best example is the transistor, which originates from semiconductor physics. The lack of a bandgap in graphene results in graphene transistors to underperform compared to their semiconductor counterparts, due to the lack of a saturation region and the lack of an off state. So, graphene transistors are not suitable for digital applications and cannot be used as amplifiers beyond few GHz, although better graphene quality and clean fabrication processes could increase the frequency up to 1 THz, especially when ballistic transport is used. On the other hand, opening a bandgap in graphene destroys its main attractive properties, such as high mobility or temperature independence of its physical parameters.

The main idea for developing nanodevices based on atomically thin materials is to use specific physical properties of 2D materials and to develop specific devices based on them. All semiconducting devices originate from semiconductor physics. Therefore, new devices should result from the new physics of atomically thin materials. For example, the ballistic transport and/or the nonlinear electromagnetic properties of graphene can be used for innovative devices in electronics and optics. Or, the large effective mass in TMDs can be exploited to reduce the detrimental effects in transistors with gate length of 5 nm and even less.

We hope that in few years, the devices based on atomically thin materials will fully exploit their unprecedented physical properties as well as the possibility to scale down electronic devices to atomic scale. The first such device—a transistor based on a single atom—was recently reported. Moreover, the ultimate computer will be based on atomic switches, which are now under development in the area of memristors and artificial learning. Thus, the adventure to atomic scale electronics has just started.

We are deeply indebted to our colleagues and friends with whom we have worked these years and have shared ideas and experience. Many thanks to Dr. George Konstantinidis and Dr. George Deligeorgis from FORTH, Heraklion, who have fabricated the first graphene devices on flakes, Prof. Dan Neculoiu, who has modeled and measured together with us the first microwave devices ever made on graphene, Dr. Antonio Radoi, who fabricated one of the first graphene photodetectors using a small bottle of graphene ink received for free from Nanointegris, Prof. Hans Hartnagel, with whom we have designed the first electronic THz graphene emitters and multipliers, Dr. Peter Blake, who provided us with the first graphene flakes and continues to do it, and to Dr. Mircea Modreanu, from whom we learned oxide physics and fabrication tricks. Many thanks also to our colleagues from the NanoRF project and FP7 project founded by EU, who shared with us novel ideas and methods.

Many thanks to many unnamed colleagues, who encouraged us to work in the area of the physics and devices based on atomically thin materials. This book, involving ten years of intense research, short holidays, and few weekends, is dedicated to all of them.

Last, but not least, we thank Dr. Claus Ascheron, who is the editor of our books since 1998. He has trusted us from the beginning, and we hope that this book will please him as have the other four books previously published at Springer in various editions.

Bucharest, Romania
Mircea Dragoman
Daniela Dragoman

Contents

About the Authors

Mircea Dragoman was born in Bucharest in 1955. He was graduated from the Polytechnical Institute in Bucharest, Electronic Faculty, in 1980. He received the doctoral degree in electronics in 1991.

Mircea Dragoman is a senior researcher I at the National Research Institute in Microtechnologies, since 1996. He is teaching since 2008 at Univ. Politehnica Bucharest, Romania, a course termed Advanced Technological Processes, which is related to nanotechnologies and advanced materials. He has realized the first carbon nanotube and graphene devices and circuits for high-frequency applications enriching the novel area of carbon-based electronics. In the period 1992–1994, he was the recipient of the Humbold Fellowship award and he has followed post-doctoral studies at Duisburg University, Germany. He was invited professor at: CNR- Istituto di Electtronica dello Stato Solido-Roma (1996), Univ. Saint-Etienne–Franta (1997), Univ. Mannheim (1998–1999, 2001–2002), Univ. Frankfurt (2003), and Univ. Darmstadt (2004); in the periods 2005–2006 and 2008–2010, he was nominated directeur de recherche at CNRS LAAS Toulouse. He has published more than 250 scientific papers in the following areas: nanoelectronics, microwaves, MEMS, and optoelectronics. He received the "Gheorghe Cartianu" award of the Romanian Academy in 1999. He is co-author of the following books: D. Dragoman, M. Dragoman "*Advanced Optoelectronic Devices*,"***Springer*** (1999), D. Dragoman, M. Dragoman, "*Optical-Characterization of Solids",**Springer*** (2002), D. Dragoman, M. Dragoman "*Quantum-Classical Analogies,"**Springer*** (2004). M. Dragoman, D. Dragoman-*Nanoelectronics. Principles and Devices, Artech House, Boston, USA*, *1st edition*, *420* pages*, (2006), second edition (2008).* D. Dragoman, M. Dragoman, "*Bionanoelectronics*", Springer 2012. D. Dragoman and M. Dragoman, "*Sheng Wu Na Mi Dian Zi Xu*" [Bionanolectronics, Chinese Edition, Science Press (2015)].

Daniela Dragoman was graduated from the University of Bucharest, Physics Faculty, in 1989, and received the Ph.D. degree from the University of Limerick, in 1993.

Daniela Dragoman is a professor at the Physics Faculty, University of Bucharest. She teaches solid-state physics and nanophysics courses at the undergraduate and postgraduate levels. Her areas of interests include the physics and applications of nanostructures, with a particular emphasis on carbon nanotubes and graphene, and modeling of quantum nanoscale devices. She was the recipient of the Alexander von Humboldt fellowship during February 1998–June 1999, and September 2001–March 2002, when she worked at the Univ. of Mannheim, Germany, and occupied the position of Directeur de Recherche at LAAS-CNRS, Toulouse, France, during July–September 2008, 2009, and 2010. She was also visiting professor at several universities in France, Germany, and Italy.

Daniela Dragoman has published more than 270 scientific papers in areas including quantum and classical optics, quantum mechanics, and nanostructures. She co-authored the books *Advanced Optoelectronic Devices* (1999), *Optical Characterization of Solids* (2002), *Quantum-Classical Analogies* (2004), and *Bionanoelectronics* (2012, with an edition in 2014 at China Science Publishing and Media Inc.) published by Springer, and *Nanoelectronics: Principles and Devices* (1st edition in 2006 and 2nd edition in 2008) published by Artech House, as well as other book chapters. She received the “Gheorghe Cartianu” award of the Romanian Academy in 1999, for the book *Advanced Optoelectronic Devices* (Springer).

Chapter 1
2D Carbon-Based Nanoelectronics

Abstract This chapter is dealing with the physics and applications of graphene in nanoelectronics, sensors and optoelectronics. Therefore, the physical properties of graphene presented in this chapter, as well as the specific phenomena encountered in this material, are directly linked to the electronic or optoelectronic devices.

1.1 Physical Properties of 2D Carbon-Based Materials

More than 10 years after the discovery of graphene (Novoselov et al. 2004), which is a single sheet of carbon atoms with a thickness of 0.34 nm arranged into a hexagonal structure, the main physical phenomena in this material seem to be well understood, although some debates are still present even today.

Graphene distinguishes itself by its uncommon physical properties, which are at the origin of unprecedented mechanical or electrical physical parameters that surpass similar characteristics of any other materials, including semiconductors. Why? The answer to this question is the subject of this section.

Graphene is formed from carbon atoms in the sp^2 hybridization state, each atom being covalently bonded to three others. Thus, graphene crystallizes in a honeycomb lattice consisting of two interpenetrated triangular sublattices, which can be visualised in Fig. 1.1. In this figure, one triangular sublattice is formed by "white" carbon atoms denoted by A, while the other consists of "black" carbon atoms denoted by B. Needless to say, the "white" and "black" carbon atoms in Fig. 1.1 are identical, the labels A and B being introduced only for visualisation purposes and for understanding the sublattices' interplay.

Graphene is at the origin of other carbon-based materials. For instance, graphite is formed from many graphene monolayers stacked as a pile, while carbon buckyballs are produced when graphene is wrapped as a sphere. On the other hand, when rolled-up, graphene forms the carbon nanotube, which is a key material for nanoelectronic devices working from few hundred of megahertz up to X rays.

Figure 1.2, which represents a side-view of the orbitals in graphene, tells us a lot about the physical properties of this material. The strong and rigid σ bonds resulted

M. Dragoman and D. Dragoman, *2D Nanoelectronics*,
NanoScience and Technology, DOI 10.1007/978-3-319-48437-2_1

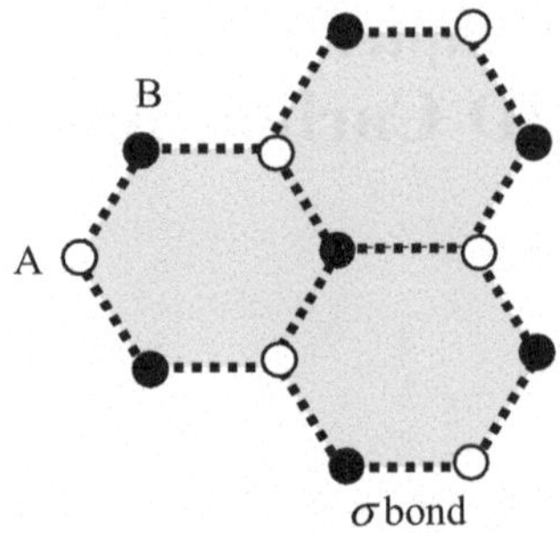

Fig. 1.1 Atomic honeycomb structure of graphene and its interpenetrated triangular sublattices

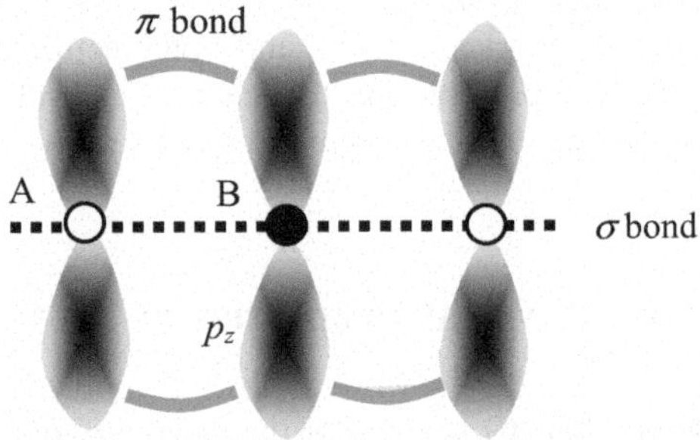

Fig. 1.2 The conduction in graphene via hybridization of p_z atomic orbitals

from the hybridization of the nearest planar sp^2 hybrid orbitals in graphene, are the source of the extraordinary strength and flexibility of this material, characterized by an exceptionally high breaking strength and Young modulus (Lee et al. 2008). However, the σ bonds are localized, and thus there are no free electrons to participate at conduction. The conduction properties of graphene are dictated only by the π and π^* bonds produced via hybridization of the remaining p_z atomic orbitals of nearest carbon atoms, denoted by A and B, as in Fig. 1.1.

Because each of the two interpenetrating triangular sublattices of the honeycomb lattice contributes to the wavefunction of charge carriers in graphene, the wavefunction is in this case a spinor rather than a scalar, as in common semiconductor materials.

The tight-binding approximation, considering only interactions between nearest neighbors, gives an energy dispersion relation of the form (Castro-Neto et al. 2009):

$$E(\boldsymbol{k}) = \pm t\sqrt{3 + 2\cos(\sqrt{3}k_y a) + 4\cos(\sqrt{3}k_y a/2)\cos(3k_x a/2)} \tag{1.1}$$

where a = 0.142 nm represents the distance between two C atoms and t = 2.75 eV is the nearest-neighbor hopping energy. A rigorous deduction of (1.1) starting from the tight-binding Hamiltonian is found in (Katsnelson 2012). The plus and minus signs in (1.1) refer to the upper (π^*) and lower (π) energy bands, the charge carriers in these bands being considered as electron and hole states. The dispersion relation of graphene is depicted in Fig. 1.3. From this figure it can be seen that at the

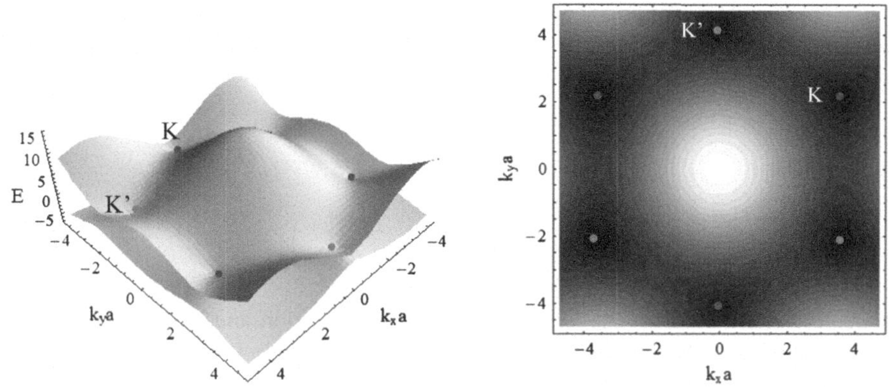

Fig. 1.3 Dispersion relation in graphene (*left*), and its contour plot (*right*)

inequivalent points K and K′ in the first Brillouin zone, called also Dirac points, the energy bandgap vanishes. From Fig. 1.3 it follows that graphene monolayer is a zero-bandgap semiconductor.

Near the Dirac points, i.e. at low energy values, the dispersion relation is:

$$E = \pm\hbar|\boldsymbol{k}|v_{\mathrm{F}}, \tag{1.2}$$

where $v_F = 3ta/2\hbar \cong c/300$ is the Fermi velocity and $|\boldsymbol{k}| = \sqrt{k_x^2 + k_y^2}$, with $\boldsymbol{k} = (k_x, k_y)$ the wavevector of charge carriers relative to Dirac points. This linear dispersion relation is the hallmark of graphene monolayer and implies that the effective mass of charge carriers vanishes near Dirac points, unlike in common semiconductors. Another difference from common semiconductors is that the transport properties in graphene are the same for electron or hole states, since the dispersion relation is completely symmetric around the Dirac point.

In the $\boldsymbol{k} \cdot \boldsymbol{p}$ approximation, the Dirac equation for the graphene monolayer is written as (Wu et al. 2010a):

$$-i\hbar v_F \sigma \cdot \nabla\psi = (E - U)\psi \quad \text{(around the } K \text{ point)} \tag{1.3a}$$

$$-i\hbar v_F \sigma^* \cdot \nabla\psi' = (E - U)\psi' \quad \text{(around the } K' \text{ point)} \tag{1.3b}$$

where $\sigma = (\sigma_x, \sigma_y)$, $\sigma^* = (\sigma_x, -\sigma_y)$, with

$$\sigma_x = \begin{bmatrix} 0 & 1 \\ 1 & 0 \end{bmatrix}, \quad \sigma_y = \begin{bmatrix} 0 & -i \\ i & 0 \end{bmatrix}, \tag{1.4}$$

and $\psi = (\psi_A, \psi_B)$ and $\psi' = (\psi'_A, \psi'_B)$ represent the spinorial wavefunctions around the K and K′ points, respectively.

The eigenvalues and eigenvectors are obtained from (1.3a, 1.3b) (considering the potential energy $U = 0$) as

$$E_\alpha = \alpha\hbar v_F (k_x^2 + k_y^2)^{1/2} \tag{1.5}$$

and

$$\psi_{\alpha\beta}(\boldsymbol{k}) = \frac{1}{\sqrt{2}}\begin{pmatrix} \exp(-i\beta\theta/2) \\ \alpha\exp(-i\theta/2 \end{pmatrix}, \tag{1.6a}$$

respectively, where $\alpha = \pm 1$ corresponds to the conduction and valence band, and $\beta = \pm 1$ corresponds to K and K′ valley regions. The angle θ is given by

$$\theta = \tan^{-1}(k_y/k_x) \tag{1.6b}$$

The relation above implies that for both K and K′ valleys a rotation in the (k_x, k_y) plane by 2π produces a shift of the wavefunction with π, phase shift called Berry phase.

An extended version of the Dirac equation, of the form

$$\{v_F[\boldsymbol{\sigma}, \boldsymbol{p}] + mv_F^2\sigma_z\}\boldsymbol{\psi} = (E - U)\boldsymbol{\psi} \tag{1.7a}$$

where $\boldsymbol{\psi} = (\psi_A, \psi_B)^T$ is the spinor wavefunction and $\boldsymbol{p} = (p_x, p_y)$ is the momentum operator, can include interactions with the substrate via the term $\propto mv_F^2$. In this case, expressing the wavefunction components as $\psi_A = \phi_A \exp(ik_y y)$ and $\psi_B = i\phi_B \exp(ik_y y)$, we obtain the following differential equation (see Dragoman and Dragoman (2009) and the references therein):

$$d^2\phi/d\xi^2 + (\Omega^2 - \beta^2)\phi - u'/\Omega_\pm(\pm\, d\phi/d\xi - \beta\phi) = 0, \tag{1.7b}$$

where the positive and negative signs correspond to $\phi = \phi_A$ and $\phi = \phi_B$, respectively. In (1.7b) we used the notations: $\xi = x/L$, $\Omega_\pm = \varepsilon - u \pm \Delta$, $\Omega = (\Omega_+\Omega_-)^{1/2}$, $\varepsilon = EL/\hbar v_F$, $u = UL/\hbar v_F$, $u' = du/d\xi$, $\Delta = mv_F L/\hbar$, and $\beta = k_y L$. The shape of the potential U dictates different numerical solutions.

The linear dispersion relation in (1.2), which is a specific characteristic of graphene monolayers, is retrieved in physics in only one other case: that of photons propagating in vacuum. However, the linearity of the dispersion relation has radically different meanings in these cases. In graphene, it signifies that the effective mass of charge carriers, which are fermions, is zero, and hence the interaction between electrons or holes and the crystalline lattice is very weak. As a result, the charge carriers in graphene propagate without collisions, i.e. ballistically, with the velocity $v_F = 3ta/2 = 10^6$ m/s $\cong c/300$, where c is the speed of light. Thus, graphene can be modelled as a 2D gas of massless fermions. In the case of photons, which are bosons, the linear dispersion relation in vacuum $E = \hbar\omega = hc/\lambda$ tells us that photons propagate with the speed of light. Therefore, the graphene monolayer can be viewed as a slow-wave structure, in which the charge carriers propagate with a velocity much slower than c.

One of the most important properties of graphene is the electrical doping. If we transfer or deposit a graphene monolayer on a SiO_2 layer with a thickness t = 300 nm, grown thermally over a doped silicon substrate, the doped Si substrate, which has a metal-like behaviour and acts as backgate, modulates the density of charge carriers n (see Fig. 1.4) via the relation (Novoselov et al. 2004).

$$n = \varepsilon_0 \varepsilon_d V_G / te, \tag{1.8}$$

where V_G is the gate voltage and ε_0 and ε_d are the dielectric permittivities of air and SiO_2, respectively.

Graphene was discovered well before 2004 (see in this respect Geim (2012)), but Novoselov and Geim (Novoselov et al. 2004) evidenced experimentally the amazing transport properties of this material. Their work was motivated by the quest of proving the electrical field effect in a metallic transistor (Geim 2011). The graphene transistor is, however, not a metallic one since n is orders of magnitude smaller than the electron density in metals. Note that the first experimental metallic transistor was found later (Dragoman et al. 2012a), and behaves similarly to the graphene field-effect-transistor (FET) depicted in Fig. 1.4.

The backgate modulation moves up and down the Fermi energy level in graphene, which is initially situated at the Dirac point, where n = 0. The Fermi energy shift is displayed in Fig. 1.5, which shows that electrical doping is able to actually select the transport carrier type (n or p) in graphene devices. Electrical doping is similar to chemical doping, and is commonly used for changing the transport type in semiconductor devices. Depending on the position of the Fermi energy/gate voltage, the transport in graphene is ambipolar, i.e. it is performed by either electrons or holes. Moreover, unlike in other materials, the transport properties of graphene monolayers are the same for electrons and holes due to the linear dispersion relation and its symmetry with respect to the Dirac points (Dragoman and Dragoman 2014), as suggested by Fig. 1.5. Therefore, electrons and holes in

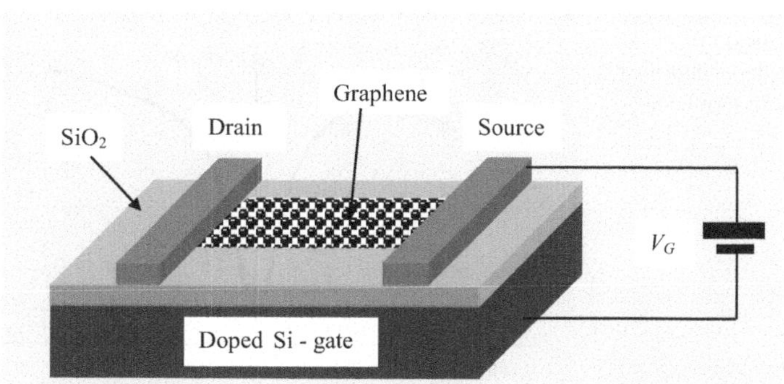

Fig. 1.4 Typical configuration of a graphene FET

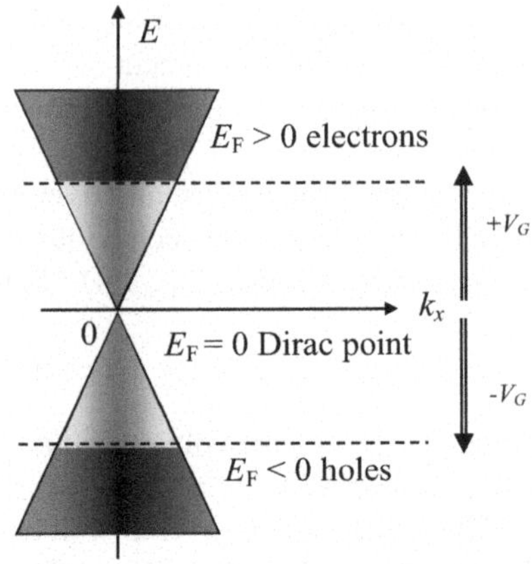

Fig. 1.5 Electrical doping in graphene

graphene are analogous to matter and antimatter particles, such as electrons and positrons, which are another groundbreaking consequence of the Dirac equations, this time for massive particles.

The conductivity dependence on the gate voltage, shown in Fig. 1.6, illustrates the similar transport properties of electrons and holes in an ambipolar monolayer graphene flake. Moreover, by modifying the gate voltage, graphene conductivity σ can be changed significantly; this behavior is similar to conductivity changes in a reversible metal-insulator transition in a large range of temperatures, including room temperature, since the conductivity in graphene depends only weakly on temperature. The dependence $\sigma(V_G)$ is essential in any graphene-based applications, and can be used to determine other parameters (Tan et al. 2006).

For instance, the mobility of charge carriers can be deduced from resistivity ρ or conductivity as:

$$\mu = (en\rho)^{-1} = \sigma/en. \tag{1.9}$$

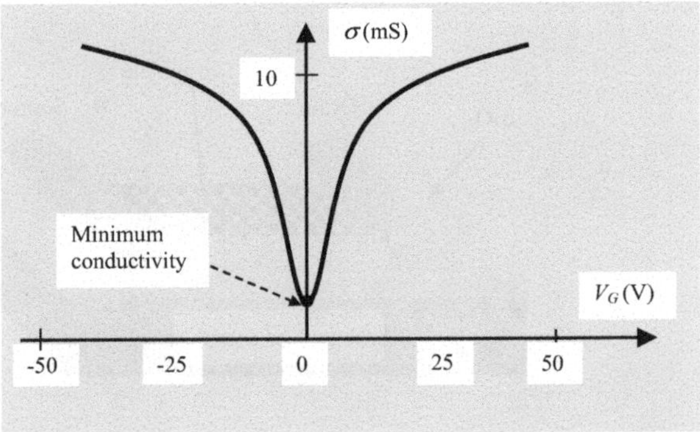

Fig. 1.6 Typical gate-voltage dependence of conductivity in graphene

In this Drude model

$$n = C_G(V_G - V_{Dirac})/e = C_G V_G/e + n_{imp} \tag{1.10}$$

where C_G is the experimentally deduced or calculated gate capacitance and n_{imp} is the carrier density associated to impurities. In other words, expression (1.10) implies that impurities induce an offset voltage, given by $V_{Dirac} = -en_{imp}/C_G$; in the absence of impurities, $V_{Dirac} = 0\,\text{V}$. The impurity concentration can be determined from V_{Dirac}, which is the voltage where the conductivity is minimum. Typical impurity concentrations in graphene samples are around 1–3 × 10^{11} cm^{-2}. From (1.9) and (1.10) it follows that the mobility in graphene is divergent near $V_G = V_{Dirac}$.

Other parameters linked to graphene conductivity are the scattering rate, given by

$$\tau = \mu E_F/(e^2 v_F) = \hbar\sigma(V_G)(\pi/n)^{1/2}/(e^2 v_F) \tag{1.11a}$$

and the mean-free-path, defined as

$$l = v_F\tau = (h/2e)\mu(n/\pi)^{1/2} \tag{1.11b}$$

Both mobility and mean-free-path have impressive values in graphene, supporting the extensive research for graphene-based nanoelectronic devices. Even in the first published paper about this material (Novoselov et al. 2004) the mobility was as high as 10,000 cm^2/V s for n = 5 × 10^{12} cm^{-2} with a room-temperature mean-free-path value of 400 nm. Two years later, it was reported that mobility values vary very much for the same carrier concentration, depending on the sample quality (Tan et al. 2006). For example, at n = 2 × 10^{12} cm^{-2} the mobility on five samples took values from 2,000 up to 20,000 cm^2/V s at low temperatures. The corresponding variation of the scattering time, from 50 to 500 fs, indicated that sample quality, i.e. a low or large defect concentration, dictates the mobility value.

In addition, the mobility in graphene is strongly dependent also on the substrate on which graphene is deposited. In the above mentioned first paper on electrical properties of graphene (Novoselov et al. 2004), the substrate was SiO_2 with a thickness of about 300 nm grown over doped Si, which plays the role of a backgate. However, the suspension of graphene at 150 nm above SiO_2 deposited over Si, as shown in Fig. 1.7, drives the mobility up to 200,000 cm^2/V s in vacuum and at low temperatures of 5 K, for graphene samples annealed at 400 K (Bolotin et al. 2008). The gate voltage in this case was maintained within the range (−5 V, 5 V) to avoid the generation of too much strain in the sample.

Note that, although very large mobilities in suspended graphene, of about 100,000–200,000 cm^2/V s, are obtained only in vacuum and at low temperatures, such values are considered as a hallmark for graphene, without any specification regarding measurement conditions and temperature. The last parameter affects much more the conductivity of suspended graphene than that of graphene on

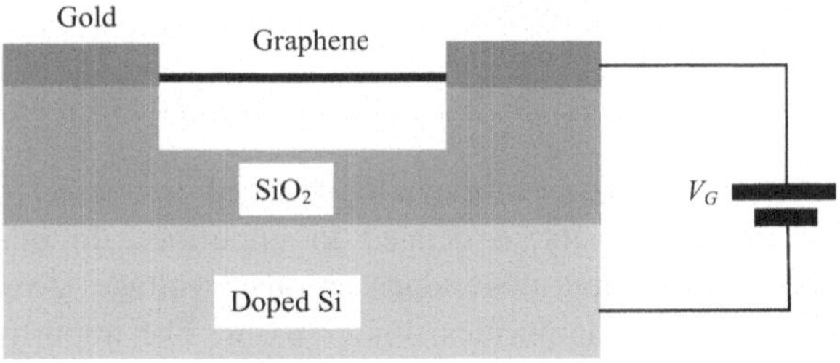

Fig. 1.7 Suspended graphene configuration

various substrates (Du et al. 2008). Moreover, suspended graphene is extremely fragile in ambient conditions and very difficult to anneal. Room-temperature ballistic transport over 1 μm and a mobility of 100,000 cm^2/V s at $n \cong 10^{11}$ cm^{-2} was obtained, nevertheless, in a graphene monolayer sandwiched between two hexagonal boron nitride (h-BN) layers with thicknesses of about 10 nm (Mayorov et al. 2011). In addition, exceptional ballistic transport, i.e. a mean-free-path of 10 μm, and a mobility exceeding 100,000 cm^2/V s was reported at room temperature in graphene nanoribbons grown epitaxially on SiC and having a width of 40 nm (Baringhaus et al. 2014). Exceptional charge transport properties of graphene transferred on (100) Ge were also demonstrated (Cavallo et al. 2014), the mobility attaining 5×10^5 cm^2/V s at 20 K and 10^3 cm^2/V s at room temperature at a carrier density of 10^{14} cm^{-2}.

Very often, the mobility of graphene is measured from electrical characteristics of a graphene FET-like configuration, whether in the form of backgate configuration, top gate configuration, or side gate geometry (see Fig. 1.8). The mobility is then extracted from the relation

$$\mu = (W/L)(\partial I_D/\partial V_D)/(en) \tag{1.12a}$$

or

$$\mu = (L/C_G W)(\partial G/\partial V_G) \tag{1.12b}$$

where W, and L are the width and the length of the channel, respectively, V_D and V_G are the drain and gate voltages, I_D is the drain current and G the channel conductance. DC measurements of a graphene FET and its geometry can be used to determine not only the mobility of graphene (Sabri et al. 2009) but of any type of channel, either organic or 2D material, such as black phosphorous (Koenig et al. 2014).

As discussed above, the mobility is dependent on the intrinsic carrier density n, which in turn is determined by the density of states (DOS) in graphene D_{gr} as follows

$$n = \int_0^\infty dE D_{gr}(E) f(E) \tag{1.13}$$

Fig. 1.8 Graphene FET with **a** backgate and top gate, and **b** side gates

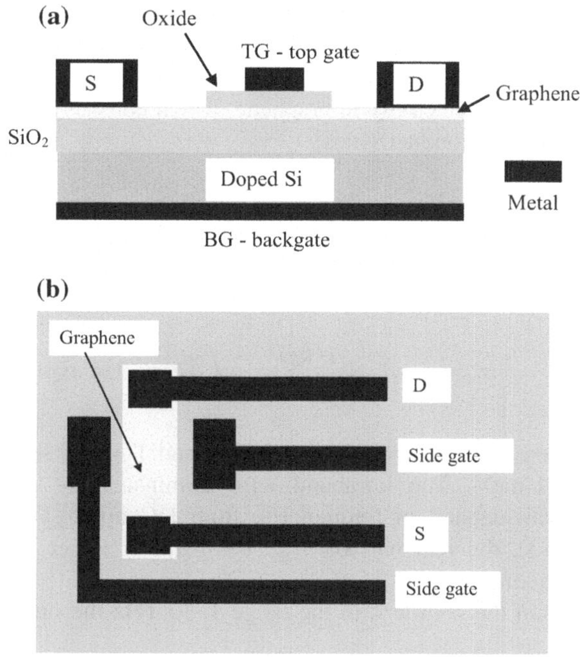

where

$$f(E) = \frac{1}{1 + \exp[(E - E_F)/k_B T]} \quad (1.14)$$

is the Fermi-Dirac distribution function. In the neighbourhood of the Dirac points, the DOS can be approximated as (Castro Neto et al. 2009)

$$D_{gr} = \frac{3\sqrt{3}a^2}{\pi} \frac{|E|}{v_F^2}. \quad (1.15)$$

At thermal equilibrium, the Fermi level is located at the Dirac point and corresponds to $E_F = 0\,\text{eV}$. The intrinsic carrier concentration in graphene is then:

$$n = p = n_i = (\pi/6)(k_B T/\hbar v_F)^2 \quad (1.16)$$

the corresponding value of this parameter at room temperature being $9 \times 10^{10}\ \text{cm}^{-2}$.

Although the DOS of graphene vanishes at the Dirac point, the conductivity has a minimum finite value at this point, as shown in Fig. 1.6. The nature of the minimum conductivity in graphene has no unique explanation, several theories attempting to elucidate this phenomenon. Among these, it is believed that the minimum conductivity $\sigma_{\min} = 4e^2/h$ originates in the quantum mechanical uncertainty principle (Dragoman 2010).

The conductivity in graphene, which consists of two components: intraband and interband conductivities:

$$\sigma = \sigma_{\text{intra}} + \sigma_{\text{inter}} \tag{1.17}$$

depends on frequency (Hanson 2008). The frequency (and temperature, T) dependence of the intraband term is given by

$$\sigma_{\text{intra}} = -i\frac{e^2 k_B T}{\pi\hbar^2(\omega - i2\Gamma)}\left(\frac{\mu_c}{k_B T} + 2\ln[\exp(-\mu_c/k_B T) + 1]\right) \tag{1.18a}$$

where μ_c is the chemical potential and Γ is the scattering rate, which is typically 0.11 meV. The intraband term dominates the graphene conductivity over an ultrawideband of frequencies, from DC up to 1 THz (Sensale-Rodríguez et al. 2013), the real part of σ_{intra} being much larger than the imaginary part in this frequency range.

On the contrary, in the range 1–30 THz the interband conductivity

$$\sigma_{\text{inter}} = -i(e^2/4\pi\hbar)\ln\left(\frac{2|\mu_c| - \hbar(\omega - i2\Gamma)}{2|\mu_c| + \hbar(\omega - 2i\Gamma)}\right) \tag{1.18b}$$

is predominant, its real part being much smaller than the imaginary part.

Graphene exhibits unusual properties when subject to a magnetic field. These properties are studied using a graphene Hall bar configuration, depicted in Fig. 1.9. For example, despite the vanishing effective mass of electrons and holes in graphene, a finite cyclotron mass of carriers can be defined as (Novoselov et al. 2005)

$$m_c = E/v_F^2 = \hbar\sqrt{\pi n}/v_F. \tag{1.19}$$

If a magnetic field B is applied normally to the graphene plane, the linear dispersion relation is modified due to the emergence of Landau levels. This is the only known method to discretize the energy spectrum in 2D graphene, the energy levels being given by (Castro-Neto et al. 2009)

$$E_n - V = \pm v_F\sqrt{2eB\hbar|n|} \tag{1.20}$$

with n an integer. The Landau levels in (1.20) are different from those for charge carriers with effective mass m_{eff} obeying the Schrödinger equation: $E_n = \hbar(eB/m_{eff})(n + 1/2)$, and result in quantized values of the Hall conductivity:

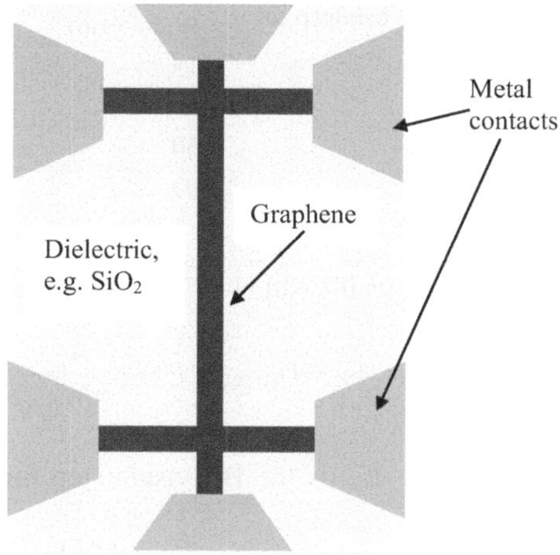

Fig. 1.9 Graphene Hall bar configuration

$$\sigma_{xy} = (n + 1/2)(4e^2/h). \tag{1.21}$$

The above quantized conductance is dependent also on the gate voltage, which changes n. The discretized energy levels in magnetic fields can be used in various applications, such as electron spectrometers (Zubarev and Dragoman 2012).

Graphene nanoribbons (GNRs) are graphene stripes with widths that do not exceed 100 nm. This spatial confinement of charge carriers in graphene changes their properties. Considering the x axis as the longitudinal axis of the GNR, the wave vector is quantized in the transverse direction, y, so that

$$k_y = n\pi/W \tag{1.22}$$

where $n = \pm 1, \pm 2, \ldots$ are integers. Thus, the dispersion relation becomes (Fang et al. 2007)

$$E_s(n, k_x) = s\hbar v_F[k_x^2 + (n\pi/W)^2]^{1/2} \tag{1.23}$$

where $s = +1$ for electrons, and $s = -1$ for holes. From (1.23) it follows that the conduction and valence bands of graphene are broken into a series of 1D subbands. More importantly, a band gap is opened with a width

$$E_g = E_{+1}(1,0) - E_{-1}(1,0) = 2\pi\hbar v_F/W, \tag{1.24}$$

which depends only on the width of the GNR. The equation above is similar to the bandgap of semiconducting carbon nanotubes (CNTs), which is inversely proportional to the nanotube diameter. The bandgap vanishes when the GNR becomes wider than 90 nm. The Table 1.1 shows some values of the bandgap extracted from experimental results (Han et al. 2007).

Table 1.1 The bandgap of GNR

Width (nm)	E_g (meV)
15	100
30	20
60	5
90	3

The DOS of the nth subband is given by

$$D_{GNR}(n, E) = \frac{4}{\pi \hbar v_F} \frac{E}{\sqrt{E^2 - E_n^2}} \mathrm{H}(E - E_n) \quad (1.25)$$

where $\mathrm{H}(E - E_n)$ is the Heaviside step function, and

$$E_n = n\pi\hbar v_F / W = nE_g/2. \quad (1.26)$$

The DOS of the GNR is then obtained as

$$D_{GNR}(E) = \sum_n D_{GNR}(n, E) \quad (1.27)$$

Similarly as in graphene, where the carrier density can be tuned via an applied gate voltage V_G, the carrier density in a GNR,

$$n_{GNR} = \frac{4k_BT}{\pi\hbar v_F} \sum_n \int_{E_n/k_BT}^{\infty} \frac{u}{\sqrt{u^2 - (E_n/k_BT)^2}} \frac{du}{1 + \exp(u - \eta)}, \quad (1.28)$$

where $\eta = E_F/k_BT$, is also dependent on V_G through the relation $E_F = eV_G$. Then, by denoting with n_i the intrinsic carrier concentration corresponding to $\eta = E_F = 0$ in (1.28), for an arbitrary Fermi level we have

$$n \cong n_i \exp(\eta), \quad p \cong n_i \exp(-\eta) \quad (1.29)$$

The total charge, defined as $Q = e(p - n)$, determines the quantum capacitance $C_q = \partial Q / \partial V_G$ as

$$C_q = [2e^2 k_B T / \pi(\hbar v_F)^2] \ln[2(1 + \cosh(eV_G/k_BT))] \quad (1.30)$$

which, when $eV_G >> k_BT$, becomes

$$C_q \cong e^2 D_{GNR}(eV_G). \quad (1.31)$$

Because the massless charge carriers in graphene are described by a Dirac-like equation, similar to that describing ultrarelativistic spin-1/2 massive particles, we can test in graphene effects predicted by quantum field theories. One of them is the

Klein paradox, which in graphene predicts that potential barriers, as those forming at an interface between regions with potential energies $V_1 > 0$ and $V_2 < 0$ (see Fig. 1.10), do not impede charge carrier transport. More precisely, charge carriers propagate through such potential barriers by (successive) transformations of electron states into hole states and vice versa (Katsnelson et al. 2006). However, the Klein paradox at a step-like barrier is more controversial since it predicts higher than 1 reflection and negative transmission coefficients, associated to the creation of electron-positron pairs at the potential discontinuity. Thus, transport experiments in *n-p* junctions in graphene are able to test these predictions, which cannot be verified up to now for relativistic massive particles. In particular, the Klein paradox at a step-like barrier with the above-mentioned meaning is invalidated in graphene, both by theory (Dragoman 2009) and experiments (Young and Kim 2009; Stander et al. 2009) demonstrating that even at a step-like barrier, i.e. at an *n-p* junction, the reflection and transmission coefficients of charge carriers are less than 1.

The Dirac spinors Ψ_1 and Ψ_2, which satisfy the Dirac-like equation for an electron wavefunction incident at an angle φ_1 with respect to the x axis from region 1 on a barrier of height $V_0 = V_2 - V_1$ and width D (see Fig. 1.11), are given by

$$\Psi_1(x,y) = \begin{cases} [\exp(ik_1x) + r\exp(-ik_1x)]\exp(ik_yy), & x \leq 0 \\ [a\exp(ik_2x) + b\exp(-ik_2x)]\exp(ik_yy), & 0 < x < D \\ t\exp(ik_3x)\exp(ik_yy), & x \geq D \end{cases} \tag{1.32a}$$

$$\Psi_2(x,y) = \begin{cases} s_1[\exp(ik_1x + i\varphi_1) - r\exp(-ik_1x - i\varphi_1)]\exp(ik_yy), & x \leq 0 \\ s_2[a\exp(ik_2x + i\varphi_2) - b\exp(-ik_2x - i\varphi_2)]\exp(ik_yy), & 0 < x < D. \\ s_3t\exp(ik_3x + i\varphi_3)\exp(ik_yy), & x \geq D \end{cases} \tag{1.32b}$$

In the above equations, the x wavevector components in the three regions can be approximated by:

$$k_1 = k_F\cos\varphi_1 = [(E - V_1)^2/\hbar^2 v_F^2 - k_y^2]^{1/2}, \tag{1.33a}$$

Fig. 1.10 Klein paradox in graphene

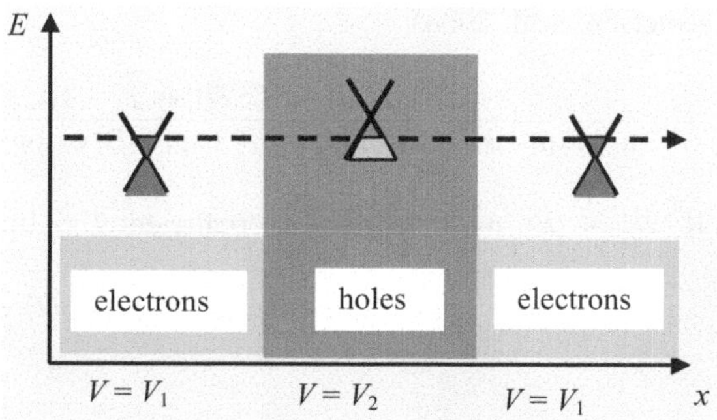

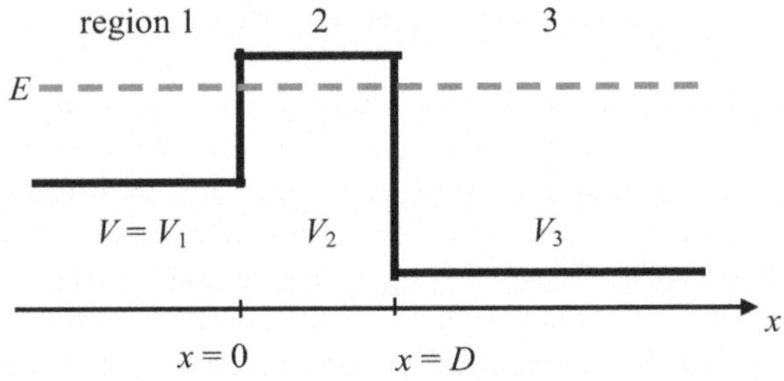

Fig. 1.11 Tunneling of Dirac fermions through a rectangular barrier

$$k_{2,3} = [(E - V_{2,3})^2/\hbar^2 v_F^2 - k_y^2]^{1/2}, \tag{1.33b}$$

where k_F is the Fermi wavenumber, $k_y = k_F \sin\varphi_1$, the propagation angles in the second and third regions are $\varphi_{2,3} = \tan^{-1}(k_y/k_{2,3})$, $s_i = \mathrm{sgn}(E - V_i)$, $i = 1, 2, 3$, with E the electron energy. Note that (1.32a, 1.32b) and (1.33a, 1.33b) can describe propagation of electrons through a gated region, which induces a potential V_0, in the presence of a bias V applied on the structure, such that the potential energy can be modelled as having a step-like variation (Dragoman and Dragoman 2007a). In the last case $V_1 = 0$, $V_2 = V_0 - eV/2$, and $V_3 = -eV$.

Then, the transmission coefficient through the barrier, defined as

$$T = s_3 \cos(\varphi_3)|t|^2/s_1 \cos(\varphi_1), \tag{1.34}$$

is computed using the wavefunction continuity requirement at the $x = 0$ and $x = D$ interfaces. The results show that, when the electron energy E is higher than the potential energy, as is the case in the first and third regions in Fig. 1.11, electrons are involved in the transport, but in the barrier region, where the electron energy is smaller than the potential energy, holes take over the charge transport role. The term "barrier" in graphene does not refer to a region of evanescent propagation as in common semiconductors, at least for normal incidence, but a region in which an abrupt change in charge carrier concentration and/or type (holes or electrons) takes place.

In particular, in a gated region in the absence of an applied bias, i.e. when $V_3 = V_1 = 0$ and $V_2 = V_0$, the reflection coefficient has a simple analytic formula (Katsnelson et al. 2006):

$$r = \frac{2i\exp(i\varphi_1)\sin(k_2 D)[\sin\varphi_1 - s_1 s_2 \sin\varphi_2]}{s_1 s_2[\exp(-ik_2 D)\cos(\varphi_1 + \varphi_2) + \exp(ik_2 D)\cos(\varphi_1 - \varphi_2)] - 2i\sin(k_x D)} \tag{1.35}$$

If $|V_0| \gg |E|$, the transmission coefficient $T = |t|^2 = 1 - |r|^2$ is

$$T \cong \cos^2\varphi_1/[1 - \cos^2(k_2 D)\sin^2\varphi_1] \tag{1.36}$$

and there are resonances, i.e. $T = 1$, when

$$k_2 D = \pi N, \tag{1.37}$$

with N an integer.

The fact that the transmission T is angle dependent was experimentally demonstrated in (Sutar et al. 2012). More importantly, the barrier becomes transparent, i.e. $T = 1$, at normal incidence, for $\varphi_1 = 0$, which is a specific manifestation of the Klein paradox, or Klein effect, in monolayer graphene. On the contrary, in bilayer graphene, in which charge carriers have a finite mass, $T = 0$ at normal incidence (Katsnelson et al. 2006).

The Klein paradox in graphene is an interesting effect from a physical point of view, but a detrimental effect for graphene ballistic devices, since the modulation of the transmission coefficient T using electrostatic gates at normal incidence is not possible because always $T = 1$. This situation, encountered only in graphene devices, is solved by using oblique incidence, i.e. oblique gates (Katsnelson et al. 2006), or magnetic fields.

In the case of an oblique electrostatic gate, as the one depicted in Fig. 1.12a, the transmission coefficient can be tuned by varying either the applied gate voltage $V_2 = eV_G$ or the applied drain voltage V_D, as can be seen from Fig. 1.12b, where the three curves correspond to different Fermi levels/gate voltages. The voltage region where $T = 0$ in this figure corresponds to the range of parameter values for which the wavenumber in the gated region is imaginary. As a result, the device displays a negative differential resistance (NDR) region, the current decreasing dramatically in the region with $T = 0$ while the voltage increases (Dragoman and Dragoman 2007a).

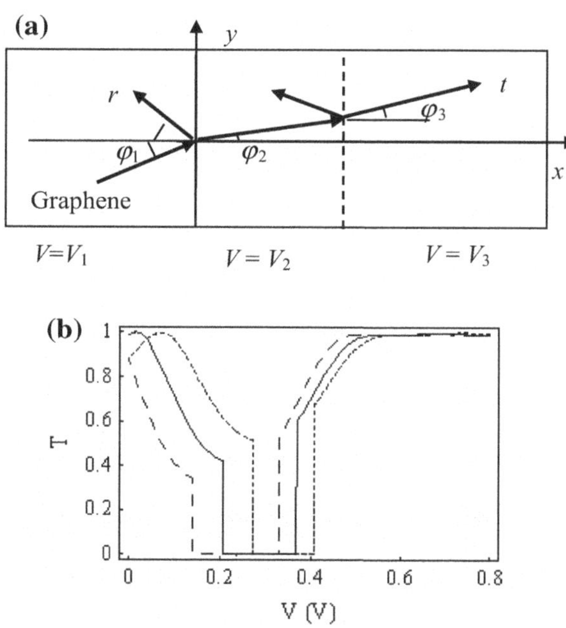

Fig. 1.12 **a** Carrier transport through an obliquely gated graphene region and **b** voltage dependence of the transmission coefficient for different gate values (from Dragoman and Dragoman 2007a)

Low transmission regions can be obtained also in periodically gated structures due to the formation of minibands (Dragoman and Dragoman 2008a).

A ballistic graphene FET with an oblique gate was recently fabricated and the NDR effect due to the modulation of the transmission was experimentally evidenced (Dragoman et al. 2014a) and is displayed in Fig. 1.13. A room-temperature peak-to-valley ratio of 9 was obtained in the NDR region of this graphene FET with a top gate having a tilting angle of 45° and a drain-source distance of 400 nm. Tens of such devices were fabricated on a chip cut from a 4-in. graphene wafer grown by CVD.

The Klein tunneling was experimentally evidenced in graphene *p-n* junctions (Young and Kim 2009; Stander et al. 2009) by tuning the top gate and backgates of devices fabricated on doped Si/SiO_2 structures. More precisely, in (Young and Kim 2009) the Klein tunneling effect is evidenced by conductance measurements under the hypothesis that the application of a magnetic field on ballistic heterostructures bends the trajectories of carriers, so that the resulting Aharonov-Bohm phase modifies the angle of incidence on the *p-n* junction. In this case, the conductance is given by

$$G \cong (8e^2/h)\sum_{k_y} |T_+|^2|T_-|^2|R_+|^2|R_-|^2 \cos\varphi \exp(-2L/l) \tag{1.38}$$

where $T_\pm$ and $R_\pm$ are the transmitted and reflected wavefunction amplitudes for the forbidden regions located at $x = \pm L/2$, φ is a sum of the semiclassical phase due to interfering trajectories and the backreflection phase, L is the gate length of the Klein device depicted in Fig. 1.14, and l is the mean-free path. In magnetic fields, the conductance oscillates as a function of the carrier density, showing phase shifts that

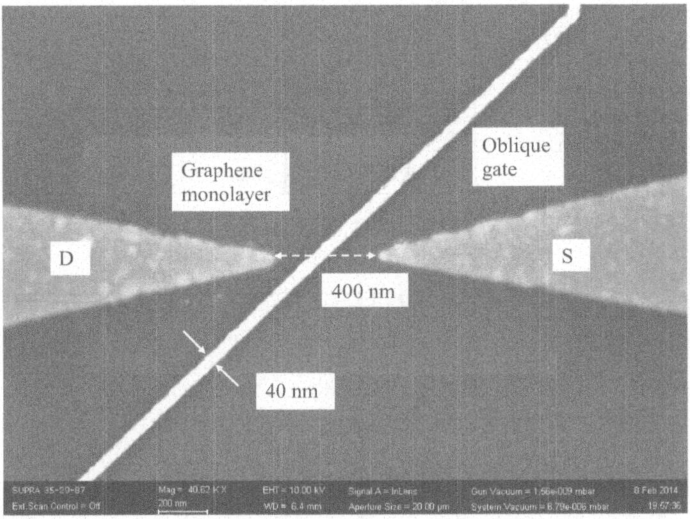

Fig. 1.13 Ballistic graphene FET with oblique gate (from Dragoman et al. 2014a)

Fig. 1.14 Ballistic device for Klein tunneling measurements

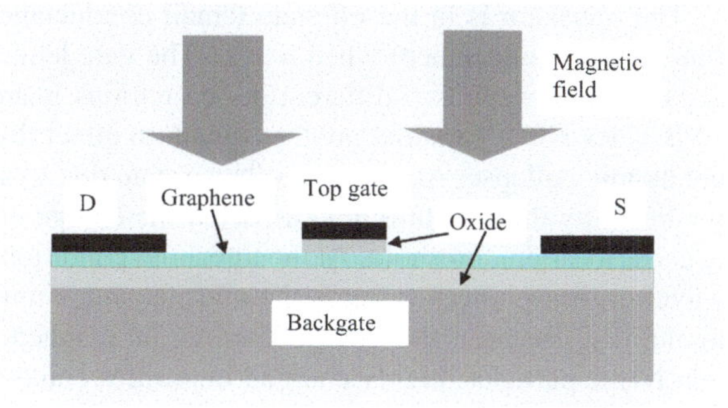

indicate perfect transmission. When $T = 1$, i.e. when Klein tunneling occurs, a phase jump of π is observed. The measurements are made at low temperature. In (Stander et al. 2009) Klein tunnelling was evidenced by resistance measurements in top-gated graphene structures with abrupt gate-induced potential steps, using a four-probe measurement setup.

A graphene Klein tunneling transistor based on a gate having a prism chain configuration was recently proposed (Wilmart et al. 2014) and is illustrated schematically in Fig. 1.15. Total reflection can be induced in this n-doped prism surrounded by p-doped regions, depending on the gate voltage. The reflection angle and the angle-dependent transmission of carriers are determined by the refractive index-like parameter

$$\nu = -(n/p)^{1/2} \tag{1.39}$$

Denoting by α_i and α_r the angles of incidence and transmission of ballistic carriers at the tunnelling prism, respectively, the Snell law applied to the prism is expressed as

$$\sin \alpha_i = \nu \sin \alpha_r \tag{1.40}$$

Fig. 1.15 Klein tunneling transistor

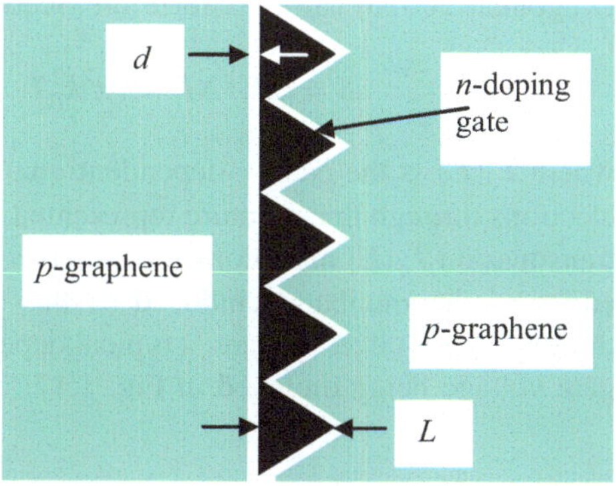

The transistor is in the off state (small conductance) when $n \gg p$ and in the on state (large conductance) when $n \cong p$. The gate length L is around 100 nm and the distance d between two different doped regions is around 10 nm.

Besides Klein paradox, another quantum effect that can be tested in graphene is the atomic collapse. An atomic collapse state due to strong Coulomb interactions is predicted by the quantum field theory if the charge of a superheavy atomic nucleus exceeds a certain threshold. At the threshold condition, i.e. at resonance, the electron wavefunction plunges towards the nucleus and, simultaneously, a positron escapes to infinity. Because the charge carriers in graphene are analoguous to massless relativistic particles, highly charged impurities could display resonances mimicking the atomic collapse. Such an experiment was recently performed (Wang et al. 2013) using artificial nuclei consisting of clusters of charged Ca dimmers placed on gated graphene devices and manipulated by a scanning tunneling microscope (STM). The atomic collapse-like states, observed in larger clusters, are distinguished by a peak in the dI/dV dependence on voltage situated below the Dirac point.

The thermal properties of graphene are also impressive, and are reviewed in (Balandin 2011). The Fourier equation for the heat flux q is $q = -K\nabla T$, where K is the thermal conductivity and ∇T is the temperature gradient. The thermal conductivity $K = K_p + K_e$ has two contributions: one from phonons, K_p, and the other from electrons, K_e. In metals, the electrons' contribution to thermal conductivity is dominant, while in carbon-based materials the phonons' contribution prevails.

Carbon nanotubes and graphene have thermal conductivities with about one order of magnitude larger than that of copper. In copper, which is used for interconnections in very large scale integrated circuits, partly because of its high thermal conductivity, K = 400 W mK^{-1}. Multiwalled CNTs have thermal conductivities >3000 W mK^{-1}, while in graphene this parameter takes values in the range 2000–5000 W mK^{-1}.

Graphene has also important thermoelectric properties. Although experiments have shown thermoelectric power values of 100 μV/K and the Seebeck coefficient S is high and tunable by the gate voltage, the figure of merit of graphene is around 1 due to a high phonon thermal conductivity K_p. The figure of merit, defined as $ZT = S^2GT/(K_e + K_p)$, with G the electronic conductance, can be enhanced in GNRs up to 4, since edge disorder can reduce K_p (Sevinçli and Cuniberti 2010). Alternatively, a periodic electrode structure patterned on graphene (Dragoman and Dragoman 2007b) can maximize the Seebeck coefficient, which is given by

$$S = \Delta V/\Delta T = (\pi^2 k_B^2 T/3e)\partial \ln T_q(E)/\partial E|_{E=E_F} \tag{1.41}$$

where $T_q(E)$ is the energy-dependent quantum transmission coefficient of ballistic electrons through the structure represented in Fig. 1.16. For maximum S values, the transmission $T_q(E)$ must consist of a series of sharp peaks, which can develop at the edges of the minibands induced in this periodic structure. At these edges, the thermoelectric effect is giant, a typical dependence of the Seebeck coefficient on the gate voltage being diplayed in Fig. 1.17.

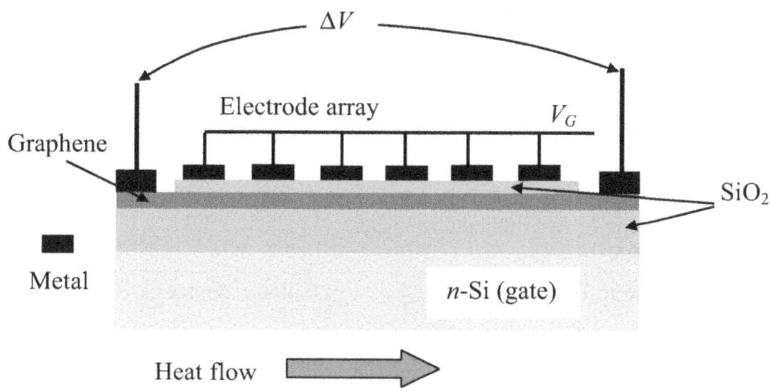

Fig. 1.16 Thermoelectric graphene structure periodically loaded with metallic electrodes

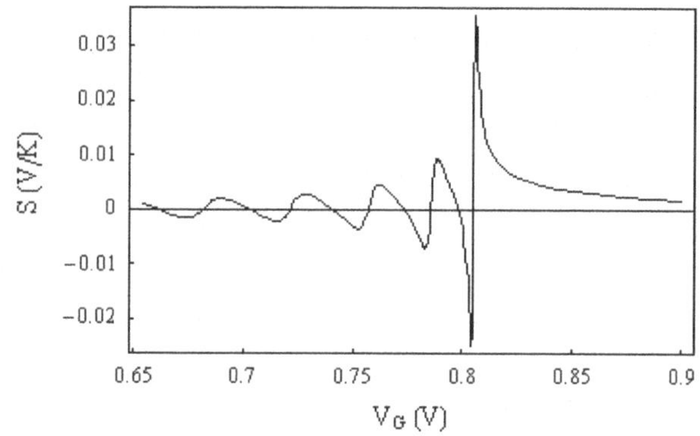

Fig. 1.17 The Seebeck coefficient of the structure represented in Fig. 1.16 (from Dragoman and Dragoman 2007b)

Finally, we summarize in Table 1.2 the main physical properties of graphene; other properties will be revealed during the next sections, which deal with graphene applications in nanoelectronics, sensors, and photonics.

The physical properties of graphene are strongly dependent on the growth methods. There are many review papers and several books about this subject (see, for instance Whitener Jr and Sheehan (2014)), the main growth methods and the yield of each of them being indicated in Table 1.3. Nowadays, two methods are intensively used. The first one, in which graphene is grown via CVD on Cu substrates, is most intensively used because it is able to produce graphene on 4-in. and even 6-in. Si wafers by transfer methods, but the quality of graphene still needs to be improved. The second method, of epitaxial graphene growth starting from SiC is also very promising, but implies very expensive equipments and high working temperatures. The quality of graphene is comparable with the CVD method, but large graphene areas are still expensive to be produced.

Table 1.2 The main physical parameters of graphene

Parameter	Value at room temperature	Applications
Mean-free-path	400 nm, even 1 μm or greater for graphene deposited on h-BN	ballistic devices: transistors and diodes for microwave, THz or photonics
Mobility	40,000 cm^2/V s (intrinsic) 100,000 cm^2/V s in suspended graphene or in graphene deposited on h-BN	transistors, high-frequency applications
Effective mass of carriers	0 close to the Dirac point	ballistic devices
Thermal conductivity	5000 W/mK, higher than in many metals	thermal interfaces, interconnects
Young modulus	1.5 TPa, ten times greater than in steel	NEMS, sensors

Table 1.3 Graphene growth methods properties

Material	Method	Yield	Graphene surface
Graphite	Repetitive peeling of HOPG	Low	Small
SiC	Desorption of Si atoms at high temperature	Moderate	Moderate (3–4 in. wafers)
GO	GO dispersion into hydrazine	High	Large
CVD	Gas mixture (CH_4 and H_2)	Very high	Very large (30 in.)

1.2 Nanoelectronics on 2D Carbon-Based Materials

1.2.1 The Graphene FET

The most studied graphene-based device is the graphene FET. All graphene physical properties recommend graphene as channel for FET applications (Dragoman and Dragoman 2009). We have presented above the top gate (TG) and backgate (BG) configurations of the graphene FET, which directly revealed the physical properties of graphene.

For a more detailed discussion of the graphene FET let us consider the TG configuration in the quantum capacitance limit regime, where the gate capacitance is much higher than the quantum capacitance: $C_G \gg C_q$. We follow further this model since it is able to provide the basic characteristics of graphene FETs. In the quantum capacitance limit regime, the drain current has the expression (Parrish and Akinwande 2012):

$$I_D = eW/L \int_0^L n(x) v_{drift} dx \qquad (1.42)$$

where L is the gate/channel length, W is the gate width, and x is the position along the channel. The drift velocity is given by

$$v_{drift} = \mu E/[1 + (\mu E/v_{sat})^{\alpha}]^{1/\alpha} \quad (1.43)$$

where $\alpha = 2$ for graphene transistors, and the saturation velocity $v_{sat} = v_F/2$ is constant.

If the channel voltage increases linearly with x:

$$V_{ch} = (x/L)V_D, \quad (1.44)$$

the surface potential is given by

$$\varphi_s(x) = V_G - V_{ch}(x). \quad (1.45)$$

Considering that

$$n = n_0 + [e\varphi_s(x)]^2/\pi\hbar^2 v_F^2 \quad (1.46)$$

where n_0 is the concentration of impurities and intrinsic carriers, we obtain the following analytic expression for the main DC characteristics $I_D(V_D, V_G)$ of the graphene FET, where V_G is the gate voltage:

$$I_D(V_D, V_G) = e(W/L)\mu^* V_D[n_0 + \gamma(V_G^2 + V_D^2/3 - V_D V_G)]. \quad (1.47)$$

In (1.47), $n_0 \cong 10^{11}$ cm^{-2}, $\mu^* = \mu[1 + (\mu V_D/v_{sat}L)]$ and $\gamma = e^2/\pi\hbar^2 v_F^2$ is a constant.

Inversely, if the mobility is not known, it can be extracted from DC measurements of $I_D(V_D, V_G)$ according to the formula

$$\mu = \frac{1}{en}\frac{\partial I_D/W}{\partial V_D/L} \quad (1.48)$$

We consider further some examples (Dragoman and Dragoman 2014). The $I_D - V_D$ curve consists of three regions, as shown in Fig. 1.18: (I) the sublinear region, (II) the saturation region, where the Dirac point is retrieved at $V_D = V_G$, and (III) the superlinear region, which originate in the lack of a graphene monolayer bandgap.

The $I_D - V_G$ dependence is represented in Fig. 1.19. Note that the drain current does not vanish when the gate voltage is zero, due to the finite minimum conductivity of graphene. The minimum value of the $I_D(V_G)$ curve is obtained at the Dirac point, $V_{G,Dirac}$. If $V_G > V_{G,Dirac}$, the conductivity is of n-type, whereas p-type conductivity is obtained for $V_G < V_{G,Dirac}$. The $I_D - V_G$ curve has a typical "V-like"-shape for any graphene FET. The ambipolar transport, which is the imprint of 2D materials, is at the origin of this V-shaped dependence.

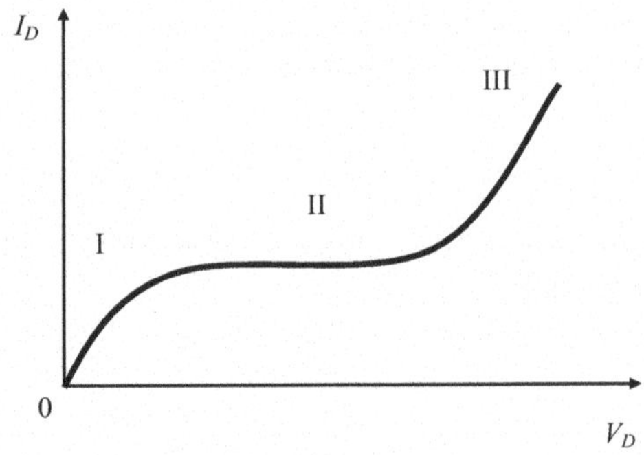

Fig. 1.18 Typical $I_D - V_D$ dependence of graphene FETs at a certain gate voltage

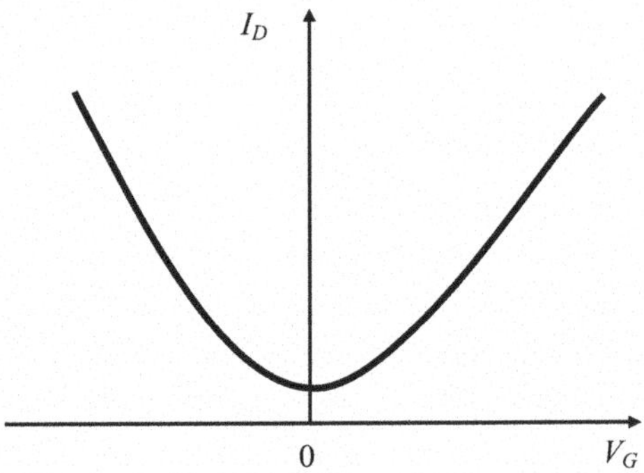

Fig. 1.19 Typical $I_D - V_G$ dependence of graphene FETs at a certain drain voltage

The large graphene/metal contact resistance is a serious problem of graphene transistors, downgrading significantly the graphene FET performances. This issue is an intensive subject of research, and a large amount of information on this subject has been gained in recent years. In particular, metals with high workfunctions, more precisely with workfunctions near that of graphene (−4.5 eV) are selected for low contact resistance (Reddy et al. 2011). For example, Ti/Pd/Au has a contact resistance of 750 Ω μm, i.e. a contact resistivity of 2×10^{-6} Ω cm^2, which is similar to the contact resistance in GaAs technology (Huang et al. 2011a). In addition, various technological treatments can reduce the contact resistance of graphene FETs (see Leong et al. (2014) and the references therein), such as soft plasma treatment, ultraviolet/ozone treatment and annealing. The role of these treatments is to destroy the resist residues, which are left on graphene FETs during their fabrication processes via lithographic techniques, such as electron-beam lithography. As shown in (Leong et al. 2014), residue-free metal (Ni)-graphene contacts can be fabricated using a shadow mask. It was found that the contact resistance is reduced from 600 to 340 Ω μm (these values depend slightly on the gate voltage) by annealing for 1 h at 300 °C, because carbon dissolves from graphene into Ni at the chemisorbed Ni-graphene interface, producing many end-contacts between metal and the dangling bonds, these contacts causing a significant decrease of the contact resistance.

There are different graphene-FETs, with different operation principles. For instance, graphene MOSFETs are similar to any MOSFETs, except that graphene is the channel. There also graphene FETs with various gate architectures, such as top gate and/or backgate, side gates, tunnel graphene FETs and vertical FETs (Schwiertz 2013).

In the graphene MOSFET (see Fig. 1.20), the top gate and/or backgate transconductance, defined as $g_m = \partial I_D/\partial V_G$, determines the cutoff frequency of the transistor $f_T = g_m/(2\pi C_G)$, where C_G is the gate capacitance.

A high cutoff frequency is possible if g_m is high and the gate length is small. The self-aligned gate technology applied to graphene FETs displays cutoff frequencies beyond 300 GHz (Liao et al. 2010). The graphene FET performances can be further enhanced if the device works in the quantum capacitance limit. This means a thickness of the oxide of few nm and high-permittivity dielectrics. For example, a graphene FET based on Y_2O_3 (relative permittivity of 10) gate dielectric with a thickness of 3.9 nm and an oxide capacitance of 2.28 $\mu F/cm^2$, works near the quantum capacitance limt (Xu et al. 2011). In this case, the cutoff frequency of the graphene FET, f_T, and f_{max} (the maximum frequency at which the transistor oscillates and provides power, i.e. the frequency for which the power gain is 1) can be defined using its equivalent circuit represented in Fig. 1.21.

For graphene MOSFETs the following intrinsic transistor formulas apply:

$$f_{T,\mathrm{int}} = g_m/[2\pi(C_{GS} + C_{GD})] \tag{1.49a}$$

$$f_{\max,\mathrm{int}} = (g_m/4\pi C_{GS}) \times (g_D R_{\mathrm{int}})^{-0.5} \tag{1.49b}$$

while the corresponding extrinsic parameters describe the whole device:

$$f_T = f_{T,\mathrm{int}}[1 + g_D(R_S + R_D) + (C_{GD} g_m (R_S + R_D)/(C_{GS} + C_{GD}))]^{-1} \tag{1.50a}$$

$$f_{\max} = (g_m/4\pi C_{GS})[g_D(R_{\mathrm{int}} + R_S + R_G) + g_m R_G (C_{GD}/C_{GS})]^{-0.5} \tag{1.50b}$$

The intrinsic graphene FET represented by the gate region and the graphene channel controlled by the gate in Fig. 1.20, has the following equivalent elements, illustrated in Fig. 1.21: the capacitance between gate and source C_{GS}, the

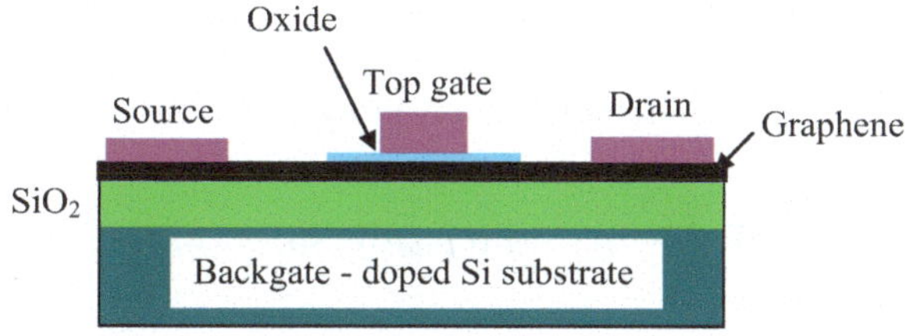

Fig. 1.20 Graphene MOSFET configuration

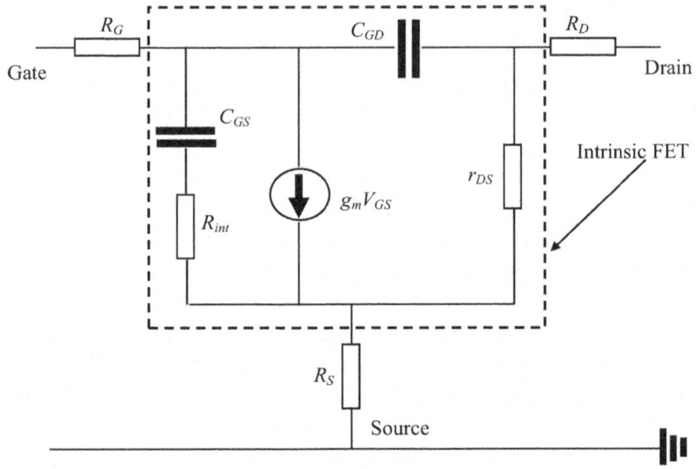

Fig. 1.21 Equivalent circuit of a graphene FET

transconductance g_m, the differential resistance r_{DS}, where $r_{DS}^{-1} = g_D = \partial I_D/\partial V_D$, and the resistance R_{int}. The gain of the intrinsic transistor is defined as $g_{\text{int}} = g_m/g_D$. On the contrary, the external transistor contains the gate, drain and source resistances, denoted by R_G, R_D and R_S, respectively. The equivalent circuit components of the intrinsic transistor, and f_T and $f_{\max}$ are bias dependent. To obtain highest values for both f_T and $f_{\max}$, the FET must work in the saturation region, condition difficult to achieve since graphene FETs do not have such a region.

There are more elaborate models of graphene FETs, valid either for diffusive transport FETs (Thiele et al. 2010) or for ballistic FETs extended to diffusive transport based on quantum mechanics (see Vincenzi et al. (2012) and the references therein). Both models are based in principle on lumped elements. A simple lumped model for graphene FETs with top and backgates, which is able to estimate the high-frequency performance of these transistors (Thiele et al. 2010), is based on the small-signal behavior in the limit $R_G = 0$. The most important parameters to be determined are C_{GS} and C_{GD}. We have (Thiele et al. 2010):

$$C_{GS} = -dQ_{ch}/dV_{G-top}\big|_{V_D=const} \tag{1.51a}$$

$$C_{GD} = -dQ_{ch}/dV_D\big|_{V_{G-top}=const} \tag{1.51b}$$

where the charge in the channel, Q_{ch}, is given by

$$Q_{ch} = eW\int_0^{V_D}(p-n)[(e\rho_{sh}\mu W/I_D) + (\mu/v_{sat})]dV \tag{1.52}$$

with the sheet carrier density ρ_{sh} as

$$e\rho_{sh} = Q_{sh} = |C_q V_{ch}/2| \tag{1.53}$$

The parameters C_q, which represents the quantum capacitance, and V_{ch}, which is the voltage across C_q, are determined from the system of equations

$$C_q = (2e^2/\pi)e|V_{ch|}|/(\hbar v_F)^2 \tag{1.54a}$$

$$V_{ch} = [V_{G-top} - V(x)](C_{ox-top}/C_{ox,eq}) + [V_{G-back} - V(x)](C_{ox-back}/C_{ox,eq}) \tag{1.54b}$$

where $C_{ox,eq} = C_{ox-top} + C_{ox-back} + C_q/2$. The system of (1.54a, 1.54b) is derived from the capacitive equivalent circuit of the gate, shown in Fig. 1.22.

Finally, the carrier density sheet can be expressed as

$$n = 2/[\pi(\hbar v_F)^2] \times \int_0^\infty [\exp[(E - E_F)/k_B T] + 1]^{-1} E dE \tag{1.55a}$$

$$p = 2/[\pi(\hbar v_F)^2] \times \int_0^\infty [\exp[(E + E_F)/k_B T] + 1]^{-1} E dE \tag{1.55b}$$

Then, C_{GS} and C_{GD} can be computed from this rather simple model, which involves a nonlinear systems of equations.

In graphene MOSFETs with long gate lengths ($L > 200$ nm) modeled as above, f_T shows a $1/L$ dependence and increases as the mobility increases. If the gate length is less than 100 nm, the cutoff frequencies of graphene FETs are very high, around 500–600 GHz, and compete with the fastest transistors ever known, such as InP HEMTs or GaAs mHEMTs. However, f_{max} of graphene MOSFETs is much lower (tens of GHz) than in InP HEMTs (hundreds of GHz). The degradation of

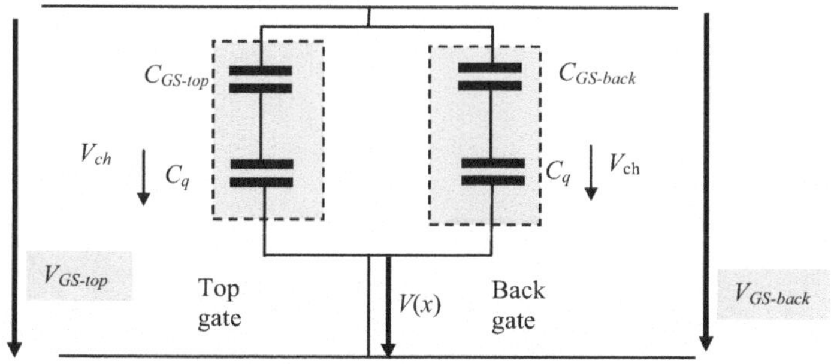

Fig. 1.22 Gate capacitances and their associated voltages

f_{max} is caused by the large drain-source conductance, which derives from the fact that saturation is not accomplished in graphene due to the lack of a bandgap. The large source-drain series resistance contributes also to the degradation of power gain and f_{max}.

In realistic transistor configurations, the graphene MOSFET has ungated regions between source/drain electrodes and the gate. These regions determine an access resistance, which lowers the drain current and the transconductance. Moreover, the ungated regions favor adsorbtion of various molecules in ambient conditions, which could degrade the transistor channel during and after processing. Therefore, for enhanced performances it is necessary to reduce to minimum the access region between source/drain and gate, expressed by an access length L_A. The best method to address the access length problem in a FET is the self-aligned source/drain electrodes technique, illustrated in Fig. 1.23. In this technique, the source and drain electrodes are defined by the gate electrode, which plays the key role in reducing L_A. The undercut region in the self-aligned transistor, shown in Fig. 1.23b, is obtained via wet-etching, which confers a very small access resistance (Jung et al. 2013). Such self-aligned FETs with a gate length of 3 μm display good performances: mobility of 6100 cm^2/V s, normalized contact resistance $R_cW = 412\,\Omega \cdot \mu m$, gate leakage currents smaller than 1 nA, and a cutoff frequency of 13 GHz. The transconductance is about one order of magnitude higher than in normal top gate graphene FETs: 114 versus 14 μS.

Better performances are obtained in FETs with sub-100 nm channel length, where the local gate is a highly doped GaN nanowire having an electrical permittivity of 10 (Liao et al. 2010). The transit time in such devices is 120–220 fs, the cutoff frequency varies in the range 700–1400 GHz, and the transconductance is about 2 mS/μm. The GaN nanowire has a sidewidth of 50–100 nm and a length of 10 μm. The electrical carrier density is about $n \cong 2 \times 10^{19}$ cm^{-3}. This graphene transistor, among the best known up to now, is represented in Fig. 1.24.

As shown in Fig. 1.25, there is a Schottky barrier between the GaN nanowire and graphene, which reduces the leakage current between the graphene channel and the GaN gate. It is worth noting that graphene has Schottky contact with many semiconductors, such as Si and GaAs, this feature being used in various applications.

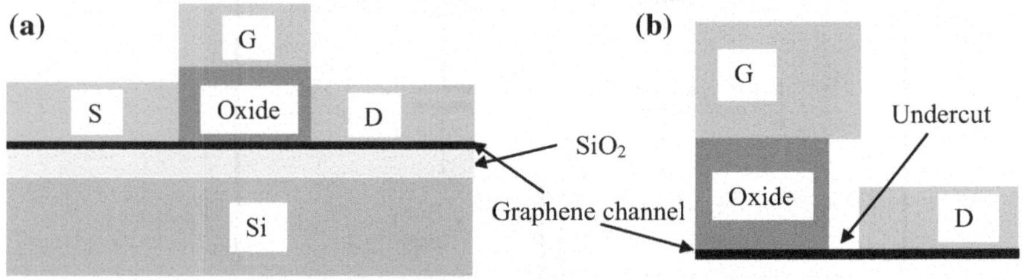

Fig. 1.23 Self-aligned graphene FET **a** cross section **b** detail in gate region

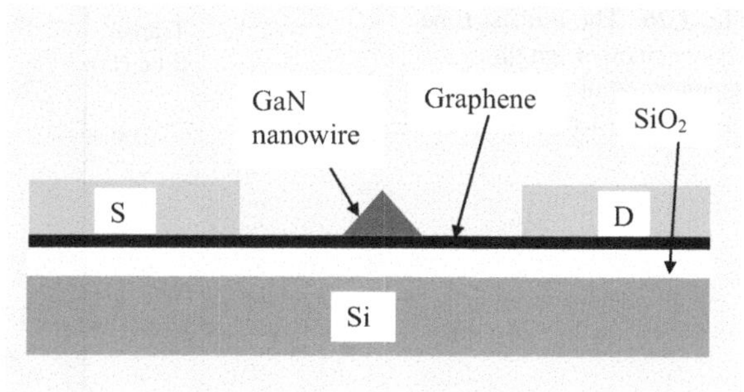

Fig. 1.24 Graphene transistor with self-aligned contacts and sub-100 nm gate length

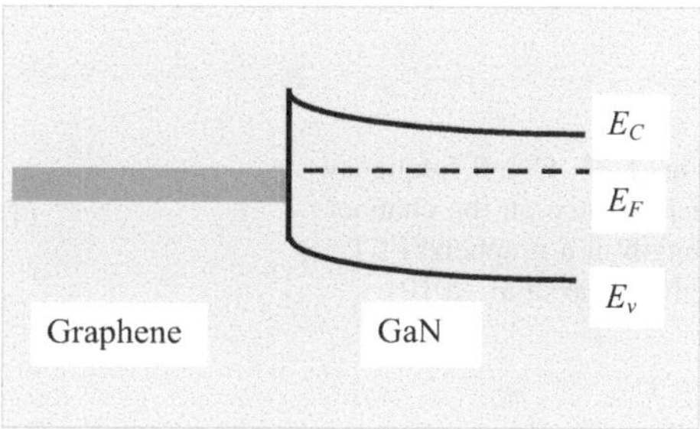

Fig. 1.25 The graphene-GaN Schottky contact

The graphene FET in Fig. 1.24 displays rather impressing properties, such as a drain current of 10 mA at $V_D = 1$ V and $V_G = 0.5$ V. The transconductance peak is −4.6 mS at the same drain voltage, i.e. 2.3 mS/μm considering that the channel has a width of 2 μm. This is among the highest transconductances attained in graphene FETs. The current density, of about 2.3×10^9 A/cm^2, is comparable with that in metallic carbon nanotubes and is three orders of magnitude higher than in CMOS transistors with copper electrodes.

The transit time of the transistor is defined as

$$\tau_{tr} = C_{GS}/g_m \tag{1.56}$$

and the intrinsic cutoff frequency is then given by

$$f_T = 1/(2\pi\tau_{tr}) \tag{1.57}$$

Considering that the gate length of the graphene FET is $L = 90$ nm, we obtain τ_{tr} about 0.19 ps at $V_D = 1$ V and a cutoff frequency of 840 GHz, meaning that this transistor is 4–5 times faster than a similar Si MOSFET with the same L, which has a cutoff frequency of only 200 GHz. The transit time of graphene FETs for various gate lengths under 100 nm is represented in Fig. 1.26, while the cutoff frequency for various gate lengths is displayed in Fig. 1.27. From Fig. 1.27 it follows that for channels shorter than 70 nm the cutoff frequencies exceed 1 THz.

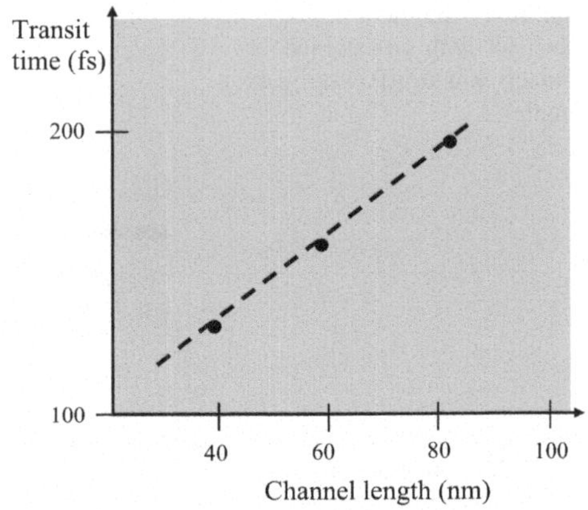

Fig. 1.26 The transist time versus channel length dependence in a graphene FET (after Liao et al. 2010)

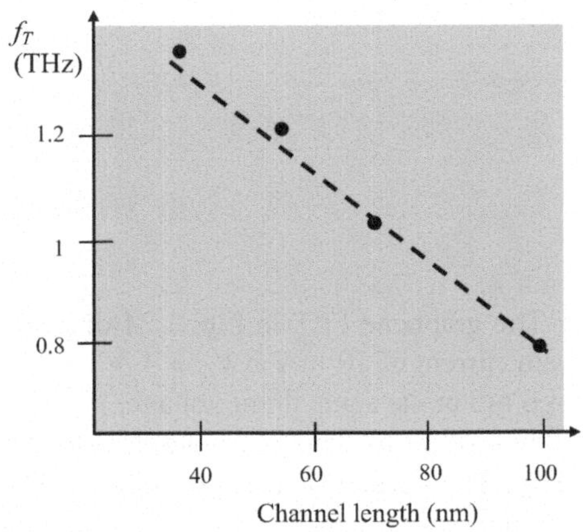

Fig. 1.27 Cutoff frequency dependence on the channel length in a graphene FET (after Liao et al. 2010)

The transit time can be also expressed as

$$\tau_{tr} = L/v_{drift} \tag{1.58}$$

the intrinsic cutoff frequency becoming then

$$f_T = v_{drift}/2\pi L \tag{1.59}$$

Because the drift velocity is about 4.3×10^7 cm/s, the transistor cutoff frequency scales as

$$f_{T,intrinsic} = 70 \text{ GHz}/L\,(\mu\text{m}) \tag{1.60}$$

High-frequency self-aligned gated transistors work up to millimeterwaves. For example, a self-aligned graphene FET fabricated via transferred gate stacks (Cheng et al. 2012) has a cutoff frequency of 427 GHz. In this case, the gate stacks are fabricated separately via e-beam patterning and RIE etching, and then transferred on any substrate on which a graphene monolayer acting as a channel is placed. The self-aligned geometry is then followed to obtain low access resistance and parasitic capacitance. The $f_{\max}$ parameter of the transistor is, however, much lower, attaining 29 GHz for a 220 nm long channel. In this case, $f_{\max}$ was determined via Mason's unilateral power gain (Cheng et al. 2012), from the relation $U^{1/2}(f) = 1$ (see Fig. 1.29), where

$$U = [|(S_{21}/S_{12}) - 1|^2/2]/(k_U|S_{21}/S_{12}| - \mathrm{Re}[S_{21}/S_{12}]) \quad (1.61)$$

and

$$k_U = (1 - |S_{11}|^2 - |S_{22}|^2 + |S_{12}S_{22} - S_{21}S_{12}|^2)/(2|S_{12}S_{22}|), \quad (1.62)$$

S_{ij}, $i, j = 1, 2$ being the scattering parameters. Figure 1.28 suggests that the highest $f_{\max}$ is obtained for an optimal gate length, and not the smallest channel length.

One of graphene FET configurations with the most impressive records is that containing an aligned T-gate on the C-face of the SiC substrate. This graphene FET configuration, schematically represented in Fig. 1.29, has $f_{\max} = 70$ GHz (Guo et al. 2013). This high $f_{\max}$ value is the result of a high mobility, of 8700 cm^2/V s, and a low contact resistance of the (Ti/Au) electrode, $R_c < 100\ \Omega\ \mu$m. The contact resistance is given by

$$R_c = (1/2)(R_{2p} - R_{channel}) \quad (1.63)$$

where R_{2p} is the two-port resistance.

The cutoff frequency of the T-gate graphene FET is 33 GHz for a gate length of 250 nm and about 100 GHz for a gate length of 100 nm.

The extrinsic $f_{\max}$ is defined by

$$f_{\max} = f_T/(2\sqrt{g_D(R_G + R_{SD}) + 2\pi f_T R_G C_G}) \quad (1.64)$$

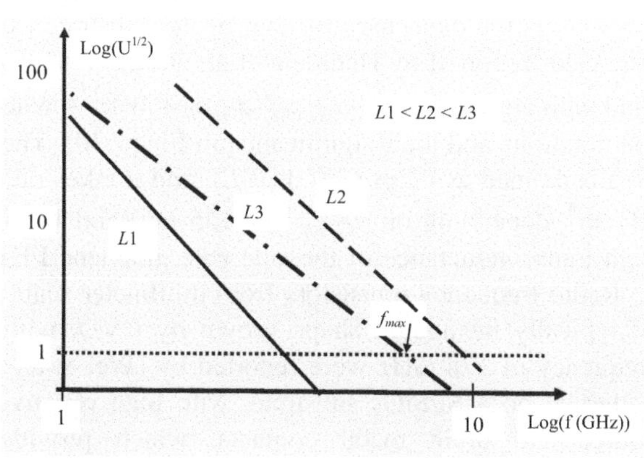

Fig. 1.28 Determination of $f_{\max}$ for self-aligned graphene FETs with different gate lengths Li, $i = 1, 2, 3$

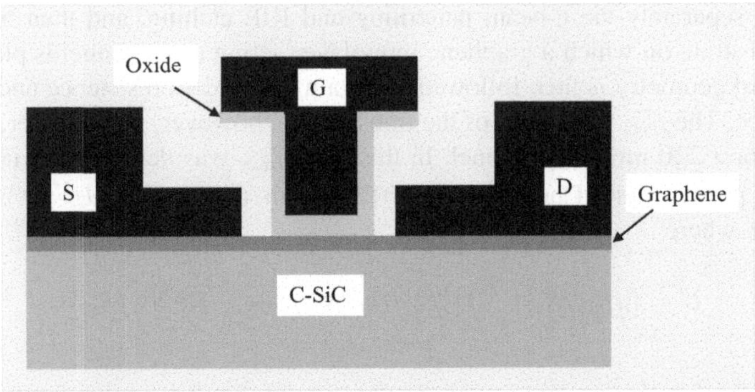

Fig. 1.29 Graphene FET with self-aligned contacts T-gate

the parameters entering expression (1.64) having the values of g_D = 2 mS, the source-drain access resistance $R_{SD} = 15\ \Omega$, $R_G = 3\ \Omega$, and $C_G = 5$ pF for the T-gate graphene FET with a 250-nm-long gate. Since in this case the resistance contact is low, we can compute $f_{\max}$ as

$$f_{\max} = f_T/(2\sqrt{g_D(R_G + R_{SD})}) \tag{1.65}$$

and it is possible to obtain simultaneously high f_T and $f_{\max}$. For instance, multifinger embedded T-shaped gate graphene RF transistors with a high $f_{\max}/f_T$ ratio were fabricated recently (Han et al. 2013). They have an $f_{\max}$ of 20 GHz, about 40% higher than f_T, at a drain bias of 1 V.

High transconductance values (590 mS/mm) can be obtained using side gate graphene transistors, as that represented in Fig. 1.30, with a lateral gate-channel separation d_{G-ch} (Hähnlein et al. 2012). In this type of graphene FETs the gate capacitance is determined by the fringing capacitance, given by

$$C_f = (\varepsilon_r + 1)\varepsilon_0\sqrt{K(1 - k^2)}/K(k) \tag{1.66}$$

where ε_r is the dielectric constant of the substrate (SiC for the graphene FET with side gates reported by Hähnlein et al. 2012), $K(k)$ is the elliptic integral of the first kind with argument $k = [d_{G-ch}/(d_{G-ch} + W)]^{1/2}$, where W is the channel width (see the notations and their significance in Fig. 1.30). The fringing capacitance per unit area is defined as $C_f' = C_f/(W \times L)$, and it takes rather small values, of 0.12–0.16 μF/cm^2, depending on d_{G-ch} (95–155 nm) and W (55–67 nm), justifying thus the high transconductance of the side gate graphene FET.

Radio frequency transistors from millimeter-scale graphene monolayer domains of typically hexagonal shape grown by CVD, with $f_{\max}$ = 35 GHz and a cutoff frequency of 178 GHz were reported by (Wei et al. 2014). These graphene FETs, patterned on a Si/SiO_2 substrate, with high resistive Si, have Pd/Au (20/40 nm) source and drain metal contacts, which provide a low contact resistance.

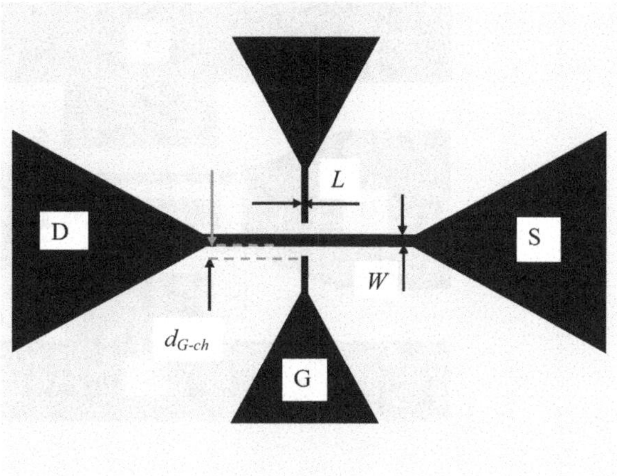

Fig. 1.30 Graphene FET with lateral gates

The 10 nm-thick Al_2O_3 gate dielectric was grown by atomic layer deposition (ALD), which is a standard growth method for all graphene FETs working at high frequencies. Initially, a 2 nm Al layer was pre-deposited on graphene by the e-beam method, in order to improve the interface between Al_2O_3 and graphene. The transistor configuration, represented in Fig. 1.31, has a fork-like gate.

In Fig. 1.32 we have illustrated the electrode layout of the graphene FET in coplanar configuration, which is the standard configuration for high-frequency transistors and integrated high-frequency transistors on various circuits, such as amplifiers, oscillators etc. The outer electrodes, i.e. the source electrodes, are grounded in the coplanar configuration, while the central electrode is the signal electrode. In order to exemplify the coplanar graphene FET configuration, Fig. 1.33 presents the SEM image of a graphene FET for high frequencies, which act as a microwave switch (Deligeorgis et al. 2010), while Fig. 1.34 shows the way in which the coplanar electrodes of this type of graphene FET are connected to probe-tips for measurements.

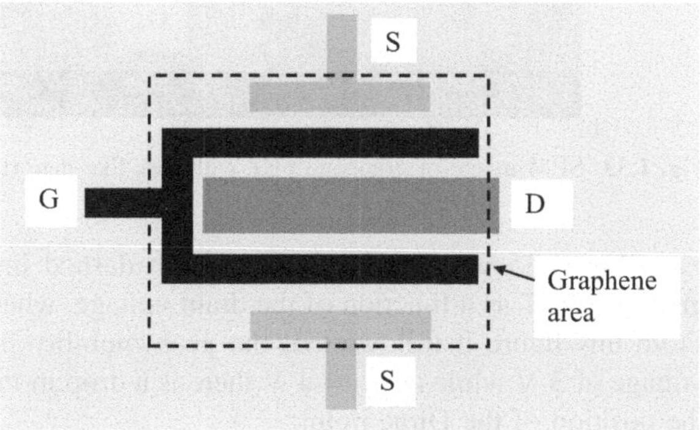

Fig. 1.31 The graphene FET with a fork-like gate

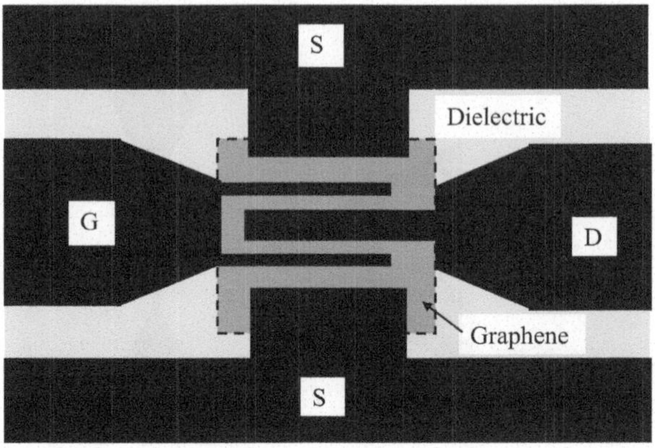

Fig. 1.32 Coplanar graphene FET electrodes

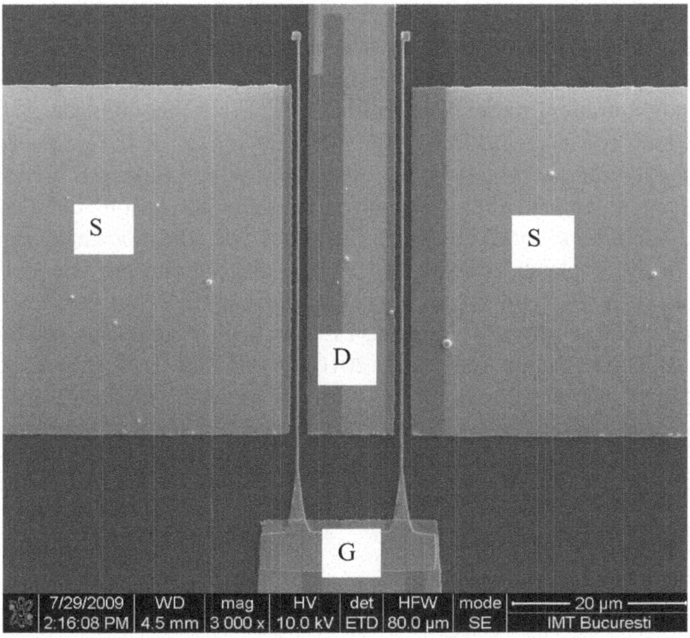

Fig. 1.33 SEM image of graphene FET with fork-like gate (from Deligeorgis et al. 2010)

In Fig. 1.35 we have represented with dashed line the mobility of a coplanar graphene FET as a function of the drain voltage, when the top gate voltage is zero. From this figure it follows that the peak mobility is of 9000 cm^2/V s at a drain voltage of 3 V while at $V_D = 4$ V there is a drop in mobility, which corresponds to the position of the Dirac point.

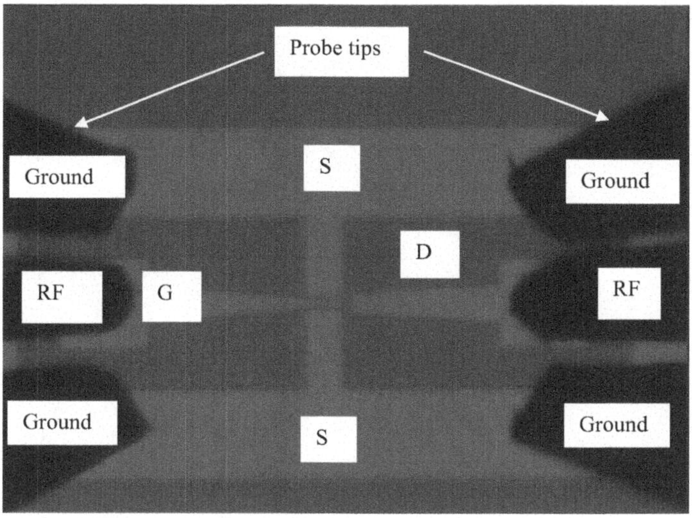

Fig. 1.34 Graphene FET in a coplanar configuration for RF measurement

Fig. 1.35 The graphene FET mobility

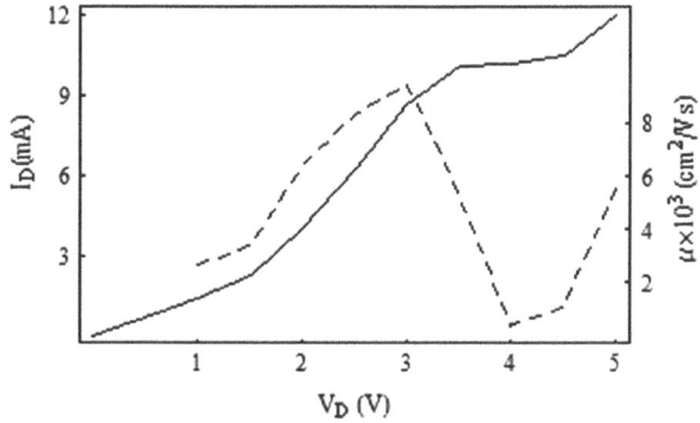

Batch production of graphene FETs on solid or flexible substrates is performed at 4- and 6-in. wafer scale, 12-in. wafers being expected in few years (Zurutuza and Marinelli 2014). An example of a graphene chip with ballistic graphene FETs is illustrated in Fig. 1.36 (Dragoman et al. 2014a).

The development of graphene FETs on flexible substrates is encouraged by the exceptional mechanical properties of graphene (Lee et al. 2008). In particular, the Young elastic modulus of monolayer graphene is 1 Tpa, conferring to this material a large bending ability, while the breaking strength is 42 N/m, graphene being stronger than steel. In this respect, it was recently demonstrated that multilayer graphene could be used as body armor against bullets (Lee et al. 2014). In addition, stretching of graphene is impressive. The stretching could attain 30% or more in graphene films on PDMS substrates, inducing a change of the sheet resistance by

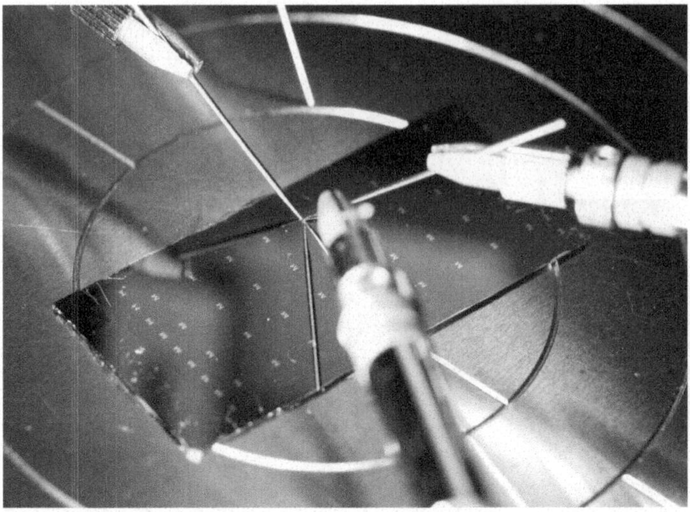

Fig. 1.36 The graphene FETs on a graphene chip grown by CVD (from Dragoman et al. 2014a)

only one order of magnitude, as can be seen from Fig. 1.37 (Kim et al. 2009), whereas the sheet resistance of ITO increases dramatically after only 2.5% stretching. The stretching in Fig. 1.37 was applied along the *y* direction, the PDMS substrate being isotropically stretched by 12% to prevent mechanical failure. Bending and stretching are the two main mechanical properties relevant for graphene FETs on flexible substrates.

There are many types of graphene FETs on flexible substrates (Sharma and Ahn 2013): with commercially available inorganic materials, with ion-gel gate dielectric for printable electronics, and/or with self-healing dielectric gates. Also, the geometry of graphene FETs depends on applications, such as digital electronics, microwave applications or sensors.

Fig. 1.37 Resistance change due to stretching of a graphene film along the *y* direction (after Kim et al. 2009)

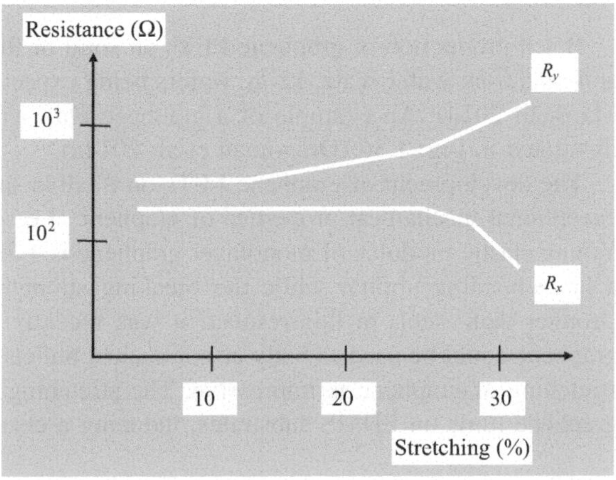

A first approach towards flexible FETs is to have a graphene channel, a high dielectric constant oxide, such as Al_2O_3, and a flexible substrate, such as PET (see Fig. 1.38). This is a self-aligned transistor on a plastic substrate, which uses the natural formation of Al_2O_3 as gate dielectric (Lu et al. 2012) and the transfer of CVD grown graphene on the PET substrate. The thickness of the oxide is at least 5 nm, and it self-heals at air exposure in a couple of hours when damaged by bending and stretching. In contrast to common gate dielectrics used for graphene FET transistors on solid substrates, such as HfO_2, which has a high growth temperature, Al_2O_3 in the example above was naturally obtained via air exposure of Al; Al is the top gate of the transistor in Fig. 1.38.

A standard approach for fabricating flexible graphene FETs is to use ion gel liquid and gelating triblock copolymer as gate dielectrics, which have a high capacitance of about 5 μF/cm^2 (Kim et al. 2010). Evidently, such organic-like transistors have low mobility, typical for organic FETs, i.e. lower than 800 cm^2/V s for holes and 190 cm^2/V s for electrons. In particular, the mobilities of the flexible device reported by (Kim et al. 2010) are about 200 cm^2/V s for holes and 90 cm^2/V s for electrons. On PET substrate, the mobility of flexible graphene FETs does not change its value by more than 20% for a bending radius varying with an order of magnitude, from 6 to 0.6 cm (see Fig. 1.39, where μ_0 is the mobility for a straight substrate). Besides mobility, the on-current of the flexible graphene FET does not show significant difference when the substrate is considerably bent.

Graphene FETs on flexible substrates display excellent performances when working as RF FETs. For example, in a bottom gate configuration, as that illustrated in Fig. 1.40, the extrinsic f_{max} reaches a maximum of 3.7 GHz while the cutoff frequency is 10.7 GHz for a strain value of 1.25% (Petrone et al. 2012), the corresponding values for a straight surface being 7.2 and 2.6 GHz. In this case, the oxide is 6 nm thick and was deposited by ALD at 150 °C on a PEN substrate. Both f_{max} and f_T are not degraded significantly up to the strain limit of 1.75%, while the mobility of the graphene FET is about 1500 cm^2/V s.

High cutoff frequencies of 25 GHz can be obtained in a multifinger embedded-gate FET fabricated on a 100-μm-thick polyimide (PI) film using as gate dielectric Al_2O_3 with a thickness of 10 nm deposited by ALD (Lee et al. 2013). The graphene channel length of this graphene FET is 500 nm and its width is 100 μm, the mobility for holes and electrons being 2800 and 3900 cm^2/V s, respectively.

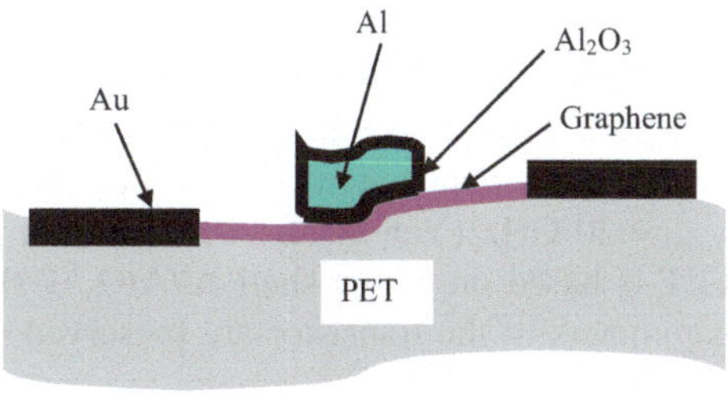

Fig. 1.38 Flexible graphene transistor with inorganic drain, source and gate

Fig. 1.39 Typical mobility-bending dependence for flexible graphene FETs

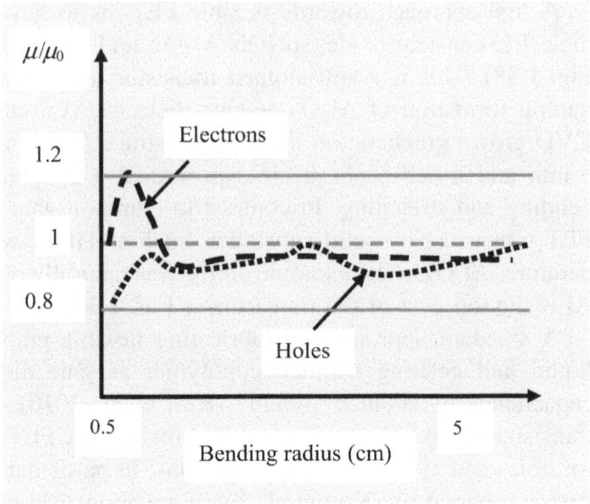

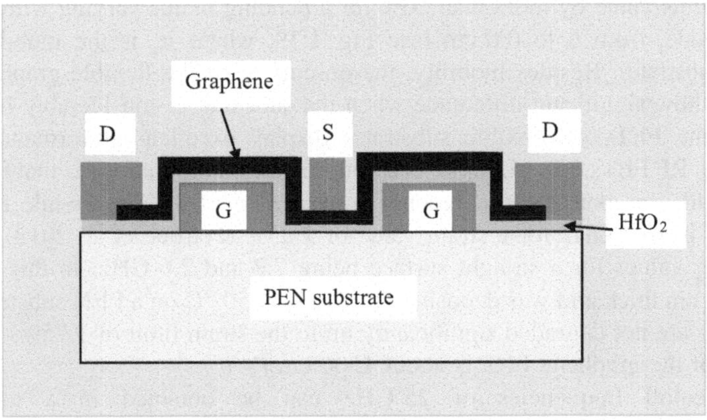

Fig. 1.40 Microwave FET on PEN substrate

The device has unchanged performances up to a bending radius of 0.7 mm, corresponding to 8.6% strain. The tensile strain is given by

$$\varepsilon = t/(2R_B) \tag{1.67}$$

where t is the substrate thickness and R_B is the bending radius.

Flexible integrated circuits on a PET substrate, such as a low-noise amplifier or a mixer, were obtained recently at an extrinsic cutoff frequency of 32 GHz and f_{max} = 20 GHz (Yeh et al. 2014). In this case, the flexible self-aligned graphene FET is based on a core-shell Al/AlO$_x$ T-shape top gate (see Fig. 1.41). The main parameters of the transistor are preserved when stretched up to 2.5%.

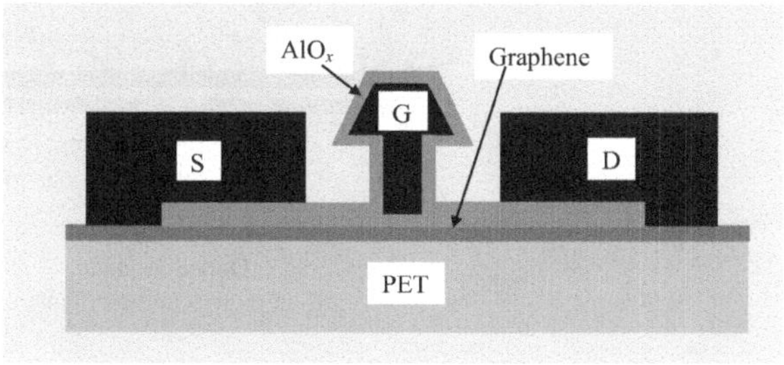

Fig. 1.41 T-shaped gate self-aligned graphene FET on PET substrate

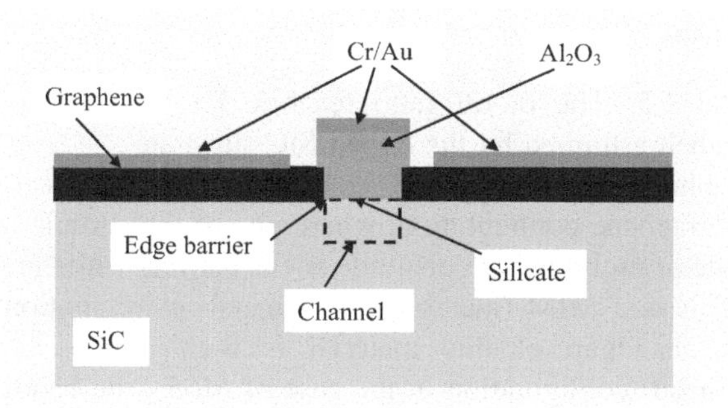

Fig. 1.42 Graphene transistor with planar edge Schottky barrier-tunneling

The ultimate performances of graphene FETs are expected from tunneling and ballistic FETs. An example of a tunneling transistor is a planar graphene/SiC FET developed for a superior on-off ratio. In this case, the flow current is horizontal over the one-dimensional tunneling barrier between multilayer epitaxial graphene, which acts as planar source and drain contacts, and the coplanar two-dimensional SiC channel (Kunc et al. 2014). The channel is an accumulation layer located at the SiC/surface silicate interface generated by high vacuum annealing, as shown in Fig. 1.42. The SiC accumulation layer and the graphene edge form a Schottky barrier, such that the equivalent circuit of the transistor consists of two back-to-back Scottky diodes controlled by the gate voltage.

At low temperatures the current is determined by tunneling, while at high temperatures the thermionic emission mechanism of the Schottky diode prevails. The drain current versus drain voltage dependence associated to thermionic emission is given by

$$I = A^* T^{3/2} W \exp(-e\phi_B/k_B T)\exp(eV/nk_B T)[1 - \exp(-eV/k_B T)] \qquad (1.68)$$

where $A^* = 0.35$ A/mK$^{-3/2}$ at room temperature and the effective barrier height ϕ_B varies between 0.8 and 0.25 eV when the gate voltage changes in the range (−4 V, 4 V). In the same gate voltage variation interval, the ideality factor n takes values

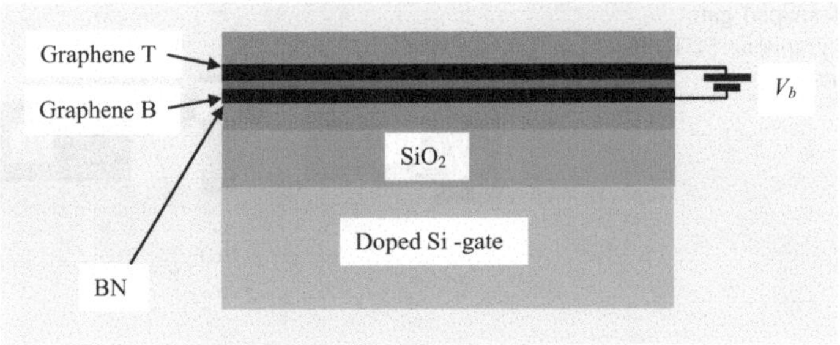

Fig. 1.43 The vertical tunneling graphene transitor

between 1.1 and 1.5. The on-off ratio reaches 5×10^6, the low mobility, of 10–20 cm^2/V s, being limited by the doped SiC substrate.

Another graphene tunneling transistor is a graphene/h-BN/graphene or graphene/MoS_2/graphene configuration, where one or few atomic layers of hexagonal boron nitride or molybdenum disulfide act as barriers while graphene are used as electrodes. This is a MIM (metal-insulator-metal) configuration in the case of h-BN since the bandgap of this material is 6 eV and is a MSM (metal-semiconductor-metal) configuration in the case of MoS_2, the bandgap of which is 1.9 eV. The graphene tunneling transistor is depicted in Fig. 1.43.

The vertical tunneling graphene FET is based on two high-quality graphene contacts, denoted as graphene T and graphene B in Fig. 1.43, encapsulated in h-BN with a thickness of 20 nm, and surrounding a 1 nm h-BN insulator layer, through which tunneling takes place. The entire structure is a tunable MIM diode with a backgate. As a result, the drain current versus drain voltage dependence is similar to a current-voltage characteristic of a diode tunable via the gate voltage, as that represented in Fig. 1.44. A high on-off ratio of 50 can be obtained in such transistors (Britnell et al. 2012). The on-off ratio reaches 10^5 if h-BN is replaced by MoS_2 in a similar vertical tunneling FET.

In the vertical tunneling graphene FET configuration, the applied electric field (in modulus) on the barrier and gate regions are given, respectively, by (Ponomarenko et al. 2013)

$$|F_b| = n_T e/\varepsilon_0 \varepsilon_r \tag{1.69a}$$

$$|F_G| = (n_B + n_T) e/\varepsilon_0 \varepsilon_r \tag{1.69b}$$

where n_B and n_T are the induced carrier densities at the two graphene electrodes and ε_r is the relative electrical permittivity of SiO_2 and h-BN, as both materials have similar dielectric constants.

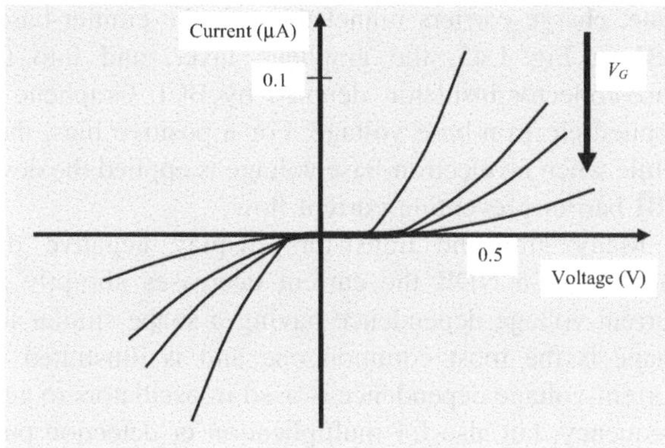

Fig. 1.44 Typical current-voltage dependence on gate voltage in a tunneling transistor

Then, the voltage between the graphene electrodes can be expressed as

$$eV_b = eF_b d - \mu(n_T) - \mu(n_B) \quad (1.70)$$

where d is the thickness of the h-BN barrier and $\mu_c(n_{T,B})$ are the chemical potentials of the corresponding graphene electrodes. The sheet density n_T under the simplified assumptions that $n_T = n_B = 0$ when no electric fields are applied and that $\mu_c(-n) = -\mu_c(n)$ can be derived from

$$e^2 dn_T/\varepsilon_0\varepsilon_r + \mu_c(n_T + \varepsilon_0\varepsilon_r E_g/e) + eV_b = 0 \quad (1.71)$$

The Landauer formula can now be used to find the current-voltage dependence.

Among other configurations of graphene FETs we mention here the vertical graphene base heterojunction transitor, which is expected to work at THz frequencies (Mehr et al. 2012). In this type of tunneling transistor, represented in Fig. 1.45, graphene is used as base control electrode for ballistic transport. In the on

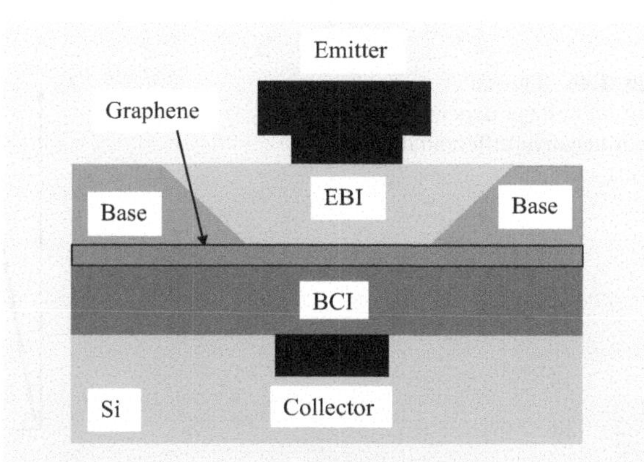

Fig. 1.45 Graphene base transistor

state, charge carriers tunnel through the emitter-base insulator layer, denoted by EBI in Fig. 1.45, the graphene layer, and into the conduction band of the base-collector insulator, denoted by BCI. Graphene tunes the EBI barrier via an applied electron-base voltage. For a positive bias, the transistor is in the on state, while when no electron-base voltage is applied the device is in the off state, the high EBI barrier preventing current flow.

Many graphene transistors display negative differential resistance (NDR) regions. In a NDR the current decreases abruptly as the voltage increases, the current-voltage dependence having a shape similar to the letters S or N. The last shape is the most common one and is illustrated in Fig. 1.46. Such nonlinear current-voltage dependence is used in oscillators to generate oscillations at a certain frequency, but also for multiplication or detection purposes.

NDR was observed in graphene/h-BN/graphene tunneling transistors, where the carriers tunnel through the h-BN insulator. The vertical graphene/h-BN/graphene structure is deposited over SiO_2/doped Si, the Si layer acting as backgate. In this case the NDR shape can be tuned by the backgate voltage. However, the peak-to-valley ratio (PVR), i.e. the ratio between the maximum and minimum currents in the NDR region, was relative small, reaching values between 2 and 4 (Britnell et al. 2013). On the contrary, the NDR effect with a PVR of 28 was observed in dual-gate graphene FETs, if V_D and V_{TG} are swept simultaneously (Liu et al. 2013). The origin of the NDR was elucidated in (Sharma et al. 2015), as the result of the competition between the field dependence of the carrier density and the drift velocity.

This is not, however, the only mechanism for generating NDR. In ballistic graphene FETs with oblique gates (Dragoman et al. 2014a), the NDR is due to an abrupt drop to zero of the transmission coefficient of carriers over a certain range of drain voltage values, V_D. This decrease in transmission produces a minimum in the $I_D - V_D$ dependence, the drain current I_D being determined by the transmission coefficient via the Landauer formula. A bias applied on the oblique gate of the ballistic graphene FET can shift the NDR region of the $I_D - V_D$ dependence, shown in Fig. 1.47, and can affect the PVR of the device; this latter parameter can be tuned up 9.

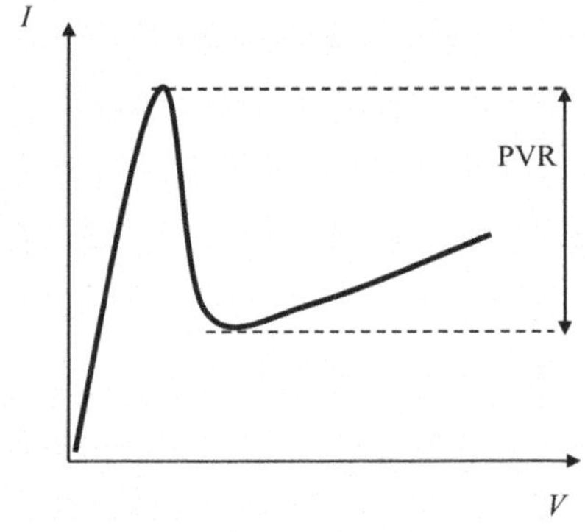

Fig. 1.46 Typical current-voltage dependence with negative differential resistance

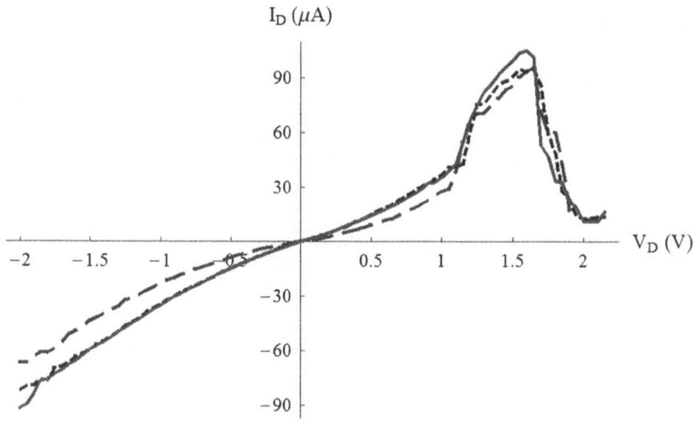

Fig. 1.47 NDR behavior in the ballistic regime of the graphene FET with oblique gate at V_{TG} = 0 V (*dashed line*), 0.5 V (*dotted line*) and 1 V (*solid line*) (from Dragoman et al. 2014a)

1.2.2 Graphene Diodes

Graphene diodes are important electronic devices, graphene widening in many ways our concepts about diodes. Diodes based on *p*-*n* junctions created by doping the host material with acceptor and donor impurities are widespread in semiconductor technology. However, it is difficult to dope graphene with impurities because some electrical properties degrade. Instead, electrical doping can be used since in graphene the carrier density and their type are modulated by a gate voltage. If the gate voltage is positive, the charge carriers are of *n* type, while *p* type carriers are generated by negative gate voltages.

An example of a *p*-*n* diode (Sutar et al. 2012) in graphene is presented in Fig. 1.48. The *p* and *n* junctions are induced by the two buried gates, made from highly doped poly-Si, and having the shape of interdigitated fingers. The diode electrodes are tilted at a certain angle since, as pointed out several times, this geometry modulates the transmission in the ballistic regime (Dragoman and Dragoman 2007a). Using suitably applied biases on the gates, all possible doping combinations, i.e. *n*-*n*, *p*-*p*, *p*-*n* and *n*-*p*, can be obtained, this versatility being impossible to be achieved in a semiconductor diode with fixed doping concentrations.

By changing the angle of incidence in this graphene diode from 0° to 30° and then to 45°, the mobility changes from 2300 to 2960 cm^2/V s, and up to 3400 cm^2/V s in various devices with the distance between gates of 150 and 200 nm, both values being smaller than the mean-free-path in graphene at room temperature. Moreover, as can be seen from Fig. 1.49, as the tilt angle increases the diode resistance is modulated by more than 100%, from 1 up to 2.5 kΩ, at $V_b = V_{G1} - V_{Dirac} = -(V_{G2} - V_{Dirac})$ = 5 V and, when the tilting angle is 45° the diode displays a particular behavior, illustrated in Fig. 1.50. At this tilting angle, as the bias voltage is varied in the interval (−20 V, 20 V), the resistance varies and

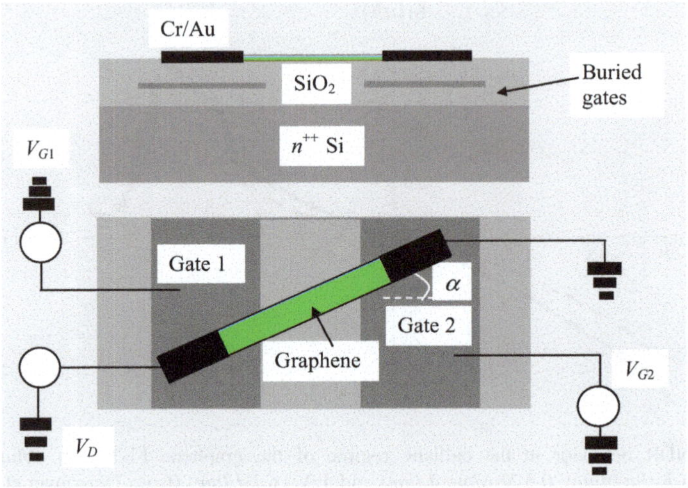

Fig. 1.48 Tilted angle *p-n* graphene junction: side view (*top*) and top view (*bottom*)

Fig. 1.49 Diode resistance as a function of the tilting angle of electrodes

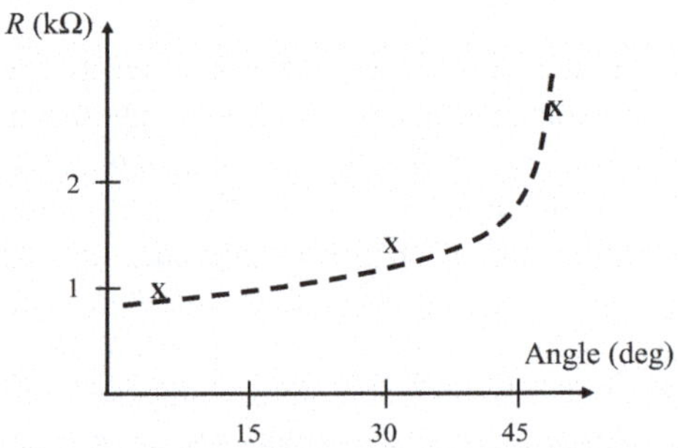

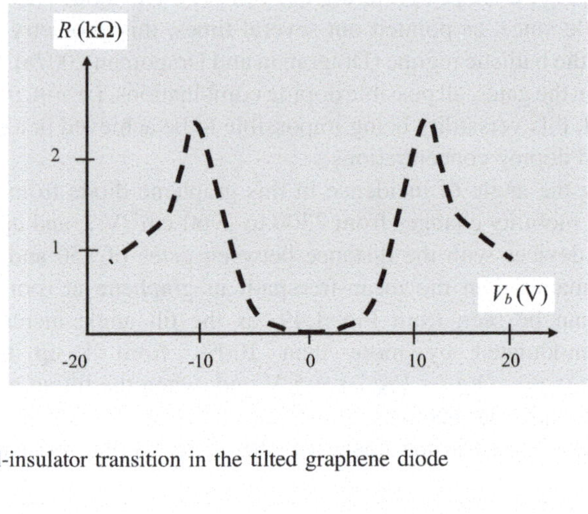

Fig. 1.50 Metal-insulator transition in the tilted graphene diode

attains a maximum value of about 2.5 kΩ. More precisely, the diode is in a low resistance state near $V_b = 0$ V, in which it behaves as a metal, and then jumps to a high resistance state at about ±10 V, in which it becomes an insulator. The on-off ratio is 2.5×10^3.

One of the most interesting graphene diodes is that resulting from the simple contact of graphene with most well known doped semiconductors, such as *n*-type Si, GaAs, 4H-SiC, and GaN. A Schottky contact forms between graphene and these semiconductors, and the charge transfer between the two materials changes the Fermi level. The first graphene-semiconductor diodes were fabricated via the transfer method, i.e. CVD grown graphene flakes were transferred on Si, GaAs, 4H-SiC, and GaN, all of *n*-type (Tongay et al. 2012). A typical configuration of a graphene-semiconductor diode is represented in Fig. 1.51. The graphene contacting areas are squares, with sides in the 500–2000 μm range. Ohmic contacts are first grown on semiconductors, then a 500–1000 nm thick SiO_2 window is grown over the semiconductor followed by gold deposition with a thickness of 500 nm. Graphene makes contact with both metal (Au) and semiconductor, showing strong rectification.

The dependence of the current density $J(T, V) = I(T, V)/A$ on temperature T and voltage V in the graphene-semiconductor diode with contact area A, through which a current $I(T, V)$ flows, is determined by thermionic emission above the Schottky barrier, and is given by

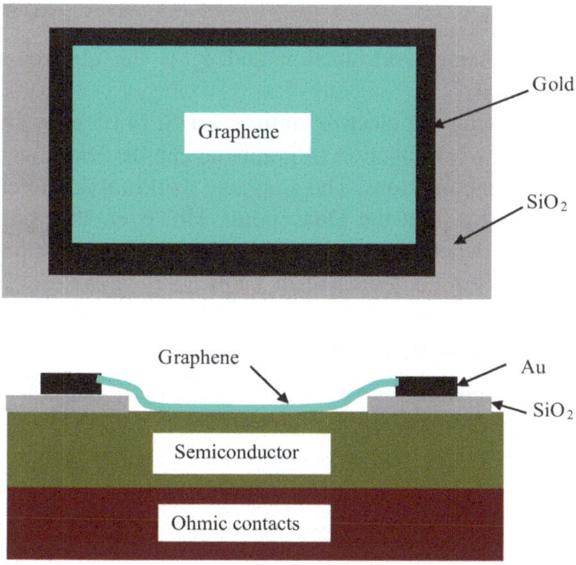

Fig. 1.51 Graphene-semiconductor diode: top view (*top*) and side view (*bottom*)

$$J(T,V) = J_s(T)[\exp(eV/\eta k_B T) - 1] \tag{1.72a}$$

where η is the ideality factor, and

$$J_s(T) = A^* T^2 \exp(-e\phi_{SBH}/k_B T) \tag{1.72b}$$

is the saturation current dependent on the zero-bias Schottky barrier height (SBH) $e\phi_{SBH}$ and the Richardson constant A^*. The ideality factor ranges between 1.2 and 5, having no association with the type of semiconductor. The imprint of the thermionic process is the linearity of the $\log(J) - V$ dependences at different T values at forward bias, as well as the linearity of the dependence $\log(J_s(T)/T^2)$ on $1/T$. From the first dependence it is possible to extract the parameters J_s and η, while from the slope of the latter curve it is possible to determine $e\phi_{SBH}$. The graphene workfunction and SBH values at different junctions are listed in Table 1.4 (Tongay et al. 2012).

Schottky barriers are described by two limits: (i) Bardeen and (ii) Schottky-Mott. In the first case, the Schottky contact is fully determined by the interface states, which pin the Femi level of the semiconductor via charge accumulation. As a result, the dependence of the barrier on the type of metal contacts is weak. The Bardeen limit is valid in the case of the graphene/GaAs diode. On the other hand, the Schottky-Mott limit is applicable for the rest of graphene/semiconductor diodes. In this case, the Schottky barrier height is given by

$$\phi_{SBH} = \Phi_{gr} - \chi_e \tag{1.73}$$

where Φ_{gr} is the graphene workfunction and χ_e is the electron affinity of the semiconducting material.

Consequently, knowing the electron affinity for Si (4.05 eV), 4H-SiC (3.4 eV) and GaN (4.1 eV), the work function of graphene can be determined as a function of the device fabrication conditions. The graphene workfunction reference is 4.6 eV when the Fermi level E_F is at the Dirac point. However, this value could differ slightly, with ±0.3 eV depending on the semiconductor, due to doping of graphene during the transfer process, as indicated in Table 1.4.

Table 1.4 Graphene/semiconductor Schottky barrier heights

Junction	Graphene workfunction (eV) (E_F at Dirac point)	Schottky barrier height (eV)	Graphene workfunction (eV) due to doping
Graphene/ *n*-GaAs	4.6	0.79	4.89
Graphene/ *n*-GaN	4.6	0.73	4.83
Graphene/ *n*-Si	4.6	0.86	4.91
Graphene/ *n*-4H-SiC	4.6	0.91	4.31

Another bilayer graphene-Si (n- or p-type) diode configuration contains the $Si/SiO_2/Si_3N_4/Cr/Au$ heterostructure depicted in Fig. 1.52 (Chen et al. 2011a). Although bilayer graphene is able to make Schottky contact with either n- or p-type Si, there are differences between these two types of diodes: (i) the graphene/n-Si diode is turning on at a positive voltage, while the graphene/p-Si diode is turning on at a negative voltage, due to the different types of carriers in the substrate, and (ii) the photocurrents are more pronounced (3–4 times higher) in the graphene/p-Si diode than in the graphene/n-Si diode at the same laser illumination power of 30 mW and at 532 nm wavelength, due to a larger space charge region.

For the graphene diode in Fig. 1.52, the Schottky barrier height is given by

$$\phi_{SBH} = (k_B T/e)\ln(A^* T^2/J_s) \tag{1.74}$$

and is equal to 0.41–0.42 eV for the bilayer graphene/n-Si diode and 0.44–0.46 eV for the bilayer graphene/p-Si diode. The ideality factor at room temperature ranges from 4.8 for the first diode type (with an n-Si substrate) up to 33 for the second one (with a p-Si substrate). However, there are almost ideal graphene/silicon Schottky diodes, which have an ideality factor of about 1.08 (Sinha and Lee 2014). To achieve such a low ideality factor in diodes in which graphene is grown by the LPCVD technique on Cu, it is necessary to etch the top 700 nm of Cu during 90s using ammonium persulfate (APS). This procedure has the role of reducing the metallic impurities in graphene. Then, graphene is transferred on the Si/SiO_2 substrate and contacted with Cr/Au contacts.

The graphene/n-GaN diode, depicted in Fig. 1.53, preserves the rectifying behavior up to 550 K, and becomes nonrectifying after 650 K. In this case GaN was n-doped with Si, the carrier concentration attaining 10^{16} cm^{-3}. After cooling, the diode shows improved rectification characteristics (Tongay et al. 2011). This extraordinary stability is due to the thermal properties of graphene, temperatures as high as 900 K being not sufficient to break the carbon-carbon bonds. The Schottky barrier height is in this case of 0.74 eV and the ideality factor is 2.9, the values of these parameters after annealing becoming 0.7 eV and 2.4, respectively.

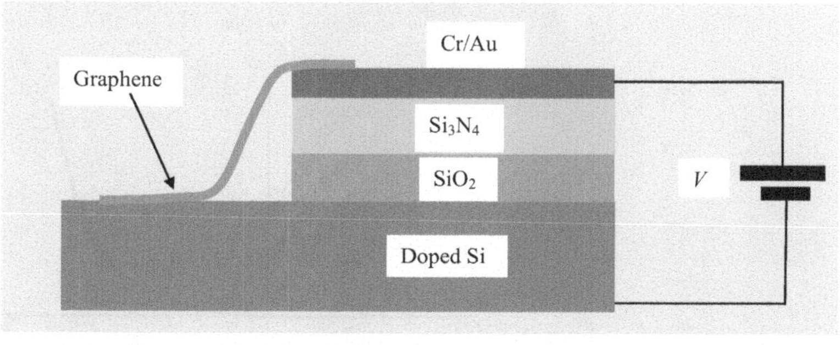

Fig. 1.52 The bilayer graphene/$Si/SiO_2/Si_3N_4/Cr/Au$ diode

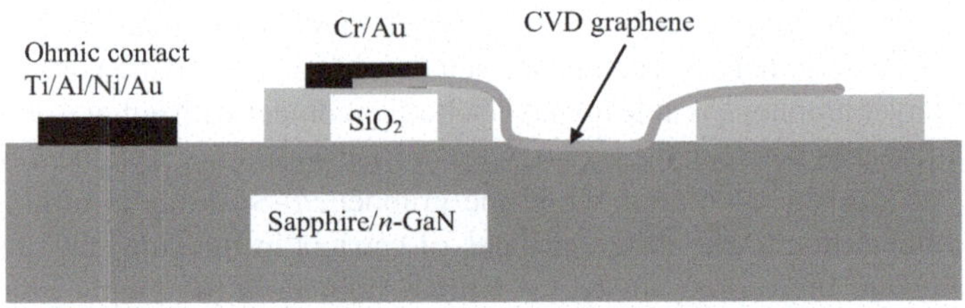

Fig. 1.53 The graphene/*n*-GaN diode

The graphene diodes presented above have DC characteristics that can be changed under the influence of various adsorbates, such as liquids and gases. Thus, chemically modulated graphene diodes can act as sensors even at elevated temperatures (Kim et al. 2013b).

Another type of graphene diodes is the self-switching diode, fabricated by carving the boundaries of a graphene strip in order to break its symmetry. This configuration is a particular example of self-switching diodes, known for more than 10 years and applied initially to 2DEG materials such as InGaAs/InAlAs (Song et al. 2003). The extension of this diode configuration to graphene is motivated by the fact that graphene is a native 2D material. A schematic representation of this diode type and its *I*–*V* characteristic are illustrated in Fig. 1.54. The self-switching diode was analyzed in several papers as a possible NDR device, but no experimental evidence of this effect was found. The voltage *V*, depending on its sign, widens or narrows the effective channel width in the 2DEG.

Fig. 1.54 Self-switching diode configuration in 2DEG (*top*) and its *I*–*V* characteristic (*bottom*)

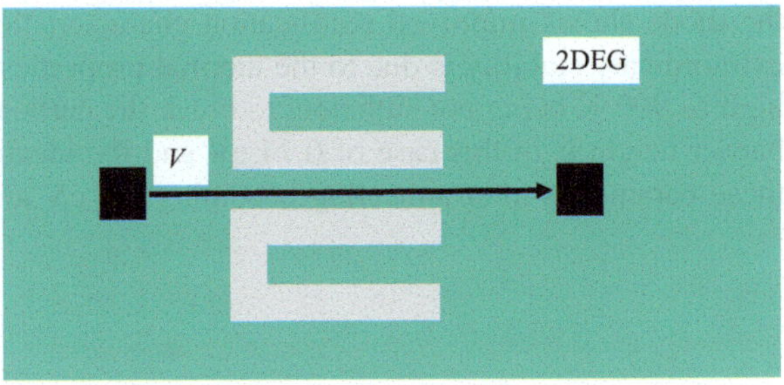

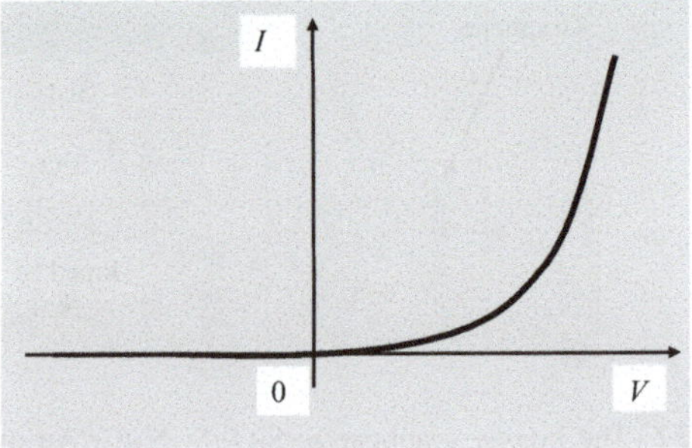

The self-switching diode is able to detect THz signals (Balocco et al. 2008). For high-frequency detection, the voltage across the diode can be modelled as

$$V = V_0 + v_0 \cos(\omega t) \tag{1.75}$$

where V_0 is the applied DC bias, and v_0 is the amplitude of the modulating signal with frequency $f = \omega/2\pi$. The current can then be expanded in a Taylor series as

$$\begin{aligned} I = I_0 + v_0 G_d \cos \omega t + v_0^2 G_d' \cos^2 \omega t + \cdots = I_0 + (v_0^2/2) G_d' \cos \omega t \\ + (v_0^2/4) G_d' \cos 2\omega t + \cdots \end{aligned} \tag{1.76}$$

where I_0 is the bias current, $G_d = dI/dV|_{V=V_0}$ denotes the differential conductance and $G_d' = d^2I/dV^2|_{V=V_0}$. The rectified DC current is then given by

$$I_r \propto (v_0^2/4) G_d' \tag{1.77}$$

and, in an open-circuit configuration, the detected voltage can be written as

$$\beta_V = (v_0^2/4) G_d'/G_d \tag{1.78}$$

Graphene self-switching diodes used as zero-bias microwave detectors were recently fabricated on graphene epitaxially grown on SiC (Westlund et al. 2015). Since the graphene channel is analogous to a lateral nanowire transistor, nine graphene channels in parallel were used to reduce the resistance and the noise-equivalent power (NEP), which is 2.2 nW/Hz$^{1/2}$ (see Fig. 1.55). The responsivity of the diode is 3.9 V/W, and detects signals from 1 to 67 GHz.

Another type of graphene diodes is the ballistic diode. A schematic representation of a ballistic Schottky diode is shown in Fig. 1.56 (Dragoman et al. 2010). It consists of two electrodes separated by a distance L, which must be less than the mean-free-path of carriers in graphene at room temperature, and an oblique gate with tilting angle θ placed between the two diode terminals. This device is similar to the recently fabricated ballistic transistor with an oblique gate described above in this chapter. A gate voltage V_G has the role of modulating the transmission coefficient T in graphene, which is less than 1 only at oblique incidence, as can be seen

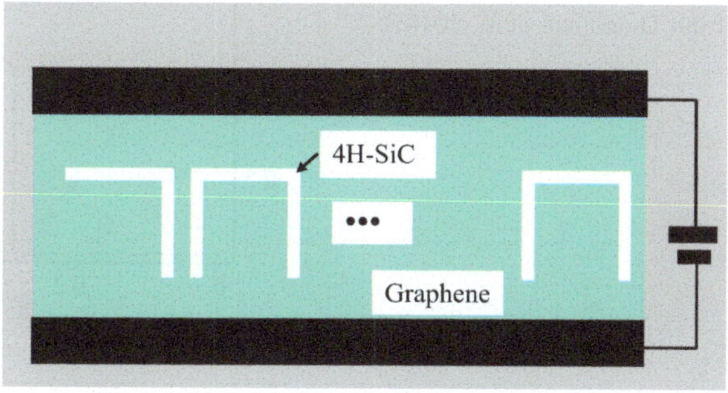

Fig. 1.55 Graphene self-switching diode RF detector

from the three curves in Fig. 1.57 corresponding to three different gate inclinations. The transmission coefficient can be calculated based on the quantum mechanical transmission-matrix approach detailed in (Dragoman and Dragoman 2007a).

Once T is known, the Landauer formula is used to compute the rectifying I–V dependences. Typical current-voltage curves corresponding to three operating points (the origin of V) chosen at 0.19 eV (solid-line), 0.29 eV (dotted-line) and 0.38 eV (dashed-line) are represented in Fig. 1.58. All three current-voltage dependences for positive applied biases can be fitted at room temperature with $I = I_S[\exp(eV/nk_BT) - 1]$. The fitting results are given in Table 1.5. Considering the Fermi velocity as $v_F = 10^6$ m/s, the traversal time between the electrodes is $t = L/v_F = 250$ fs, the resulting cutoff frequency being $f_T = 1/t = 4$ THz.

From the ballistic graphene Schottky diode example it follows that there are no reverse currents, as in the case of the semiconductor Schottky diode, and that the ideality factor of this device can be tuned up to the optimal value, equal to 1, by changing the tilting angle of the gate. What is the significance of $n < 1$? Such a value is not encountered in common semiconductors, where the Schottky rectification effect is linked to the potential energy barrier at the interface between two

Fig. 1.56 Ballistic graphene Schottky diode

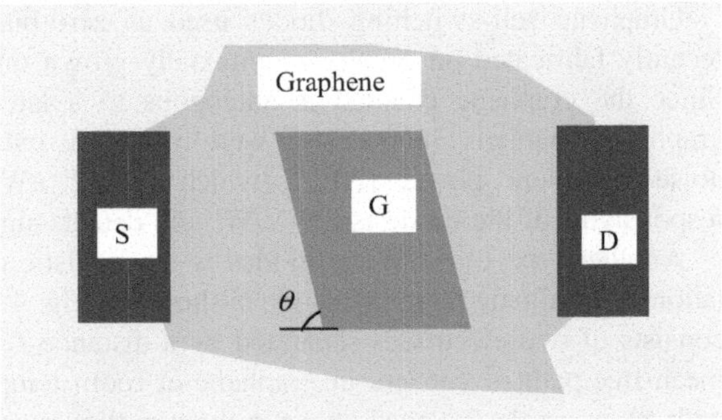

Fig. 1.57 The transmission at three gate inclination angles 20° (*solid line*), 30° (*dotted line*) and 40° (*dashed line*) (from Dragoman et al. 2010)

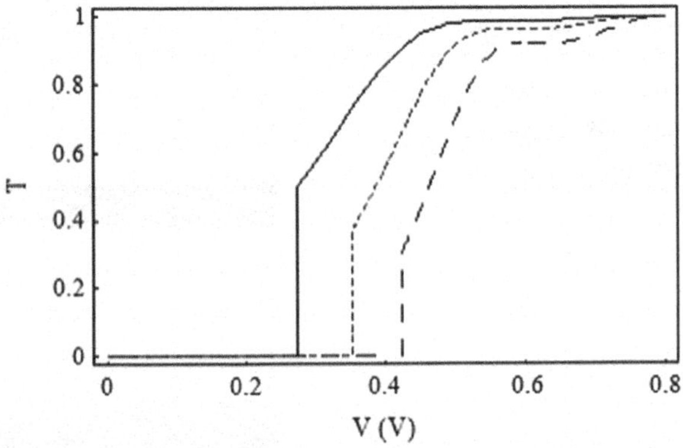

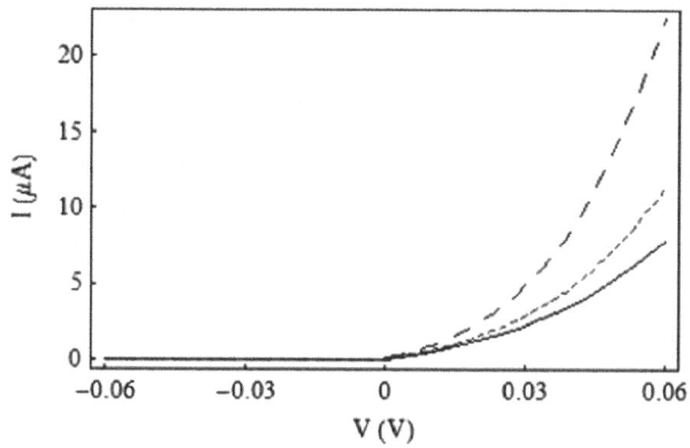

Fig. 1.58 The current-voltage dependence (from Dragoman et al. 2010)

Table 1.5 The schottky fitting values

Incident angle θ (°)	I_S (μA)	n
20	1.49	1.25
30	1.35	1.02
40	1.7	0.86

materials. In graphene, however, the mechanism is different: the rectification is the result of the lack of a bandgap in the energy dispersion relation.

Another type of rectifying ballistic diode is the geometric diode depicted in Fig. 1.59. The current-voltage dependence of this graphene diode is based also on the Landauer formula (Dragoman and Dragoman 2013). In order to compute the transmission coefficient, the diode is sectioned in a series of thin regions with a constant width, as indicated in Fig. 1.59. Then, the Dirac equation is solved in each region by imposing continuity conditions at the boundaries. It is found that the geometric diode has a certain region of width $\pi\hbar v_F/d_{out}$ where the transmission coefficient vanishes and thus the current is negligible. The current-voltage dependence of this diode is dependent on the Fermi energy, which can be tuned by a gate voltage.

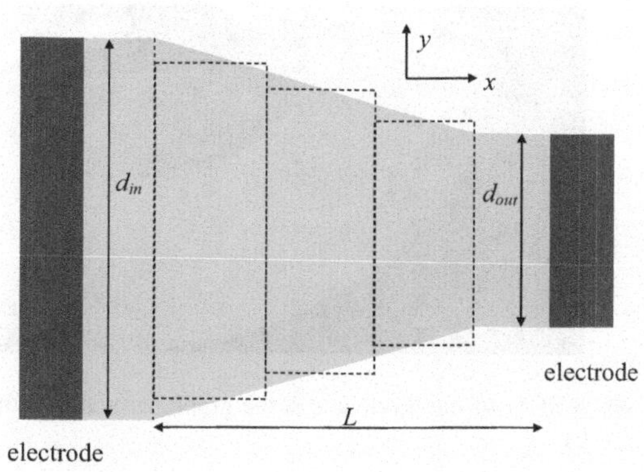

Fig. 1.59 The graphene geometric diode (from Dragoman and Dragoman 2013)

The SEM image of the diode, as well as the graphene wafer with diodes, are shown in Fig. 1.60a, b. The diode was fabricated on a 4-in. graphene wafer (Dragoman et al. 2014b), the diodes having a length of 100 nm, a shoulder d_{in} = 100 nm and a neck (d_{out}) of only 30 nm. The fabrication of the diodes involved the following steps: (i) e-beam lithography patterning of the regions without graphene monolayers, (ii) PMMA deposition, (iii) patterning of the graphene shapes with an e-beam, (iv) RIE equipment to cut the graphene in trapezoidal

(a)
(b)
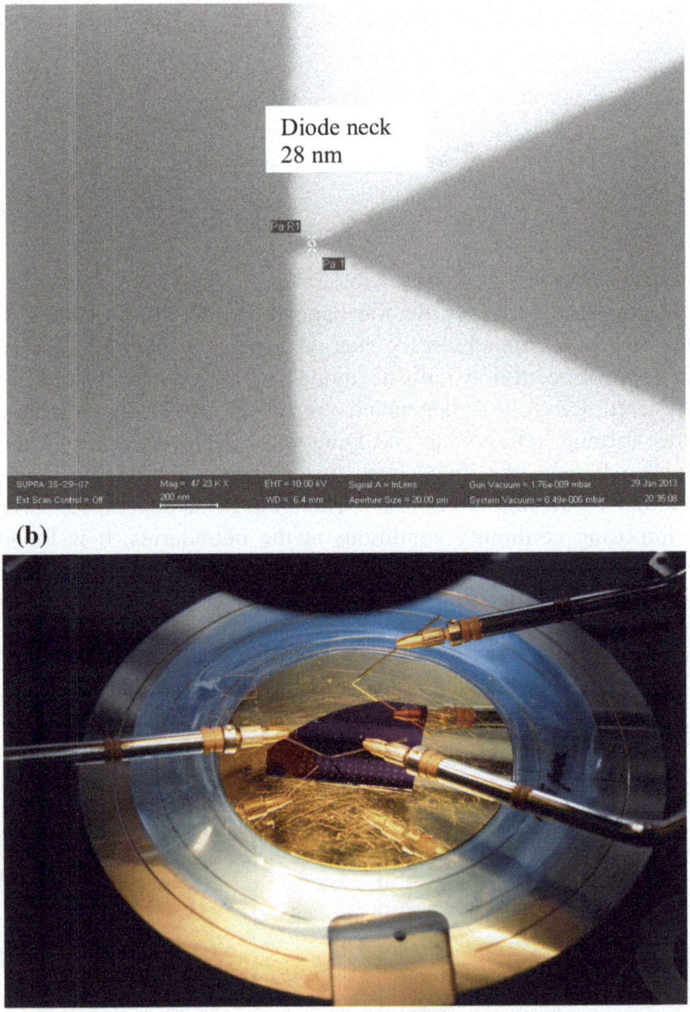

Fig. 1.60 **a** SEM of the diode and **b** the graphene wafer with geometric diodes (from Dragoman et al. 2014b)

shapes, (v) PMMA coating and e-beam lithography of metallic contacts via an e-gun evaporation chamber.

Experiments have shown that geometric diodes are able to rectify up to even 28 THz, and potentially at optical frequencies (Zhu et al. 2013b). Geometric diodes can be used as rectennas, i.e. antennas integrated with a geometric diode that rectifies in the absence of the bias signal. The rectenna is the IR analogue of a solar cell; note that the largest part of the solar energy is located in IR. The rectenna based on a geometric diode is represented in Fig. 1.61.

1.2.3 Graphene Detectors and Receivers

The simplest graphene detector is a coplanar line (CPW) patterned over a graphene monolayer flake (Dragoman et al. 2012b), as shown in Fig. 1.62. Such a graphene radio wave detector was fabricated on a high-resistivity Si substrate, on which 300 nm of SiO_2 is grown by the thermal oxidation method. Then, three parallel gold metallic electrodes forming a CPW were patterned over the graphene monolayer.

The detection of the modulated high-frequency signals is based on the nonlinear DC current-voltage characteristic of graphene with gold contacts, illustrated in Fig. 1.63 with solid line. Further, the experimental *I–V* curve is fitted with the equation

$$I = I_0[\exp(V/V_0) - 1], \tag{1.79}$$

the fitting results being marked with dashed lines in Fig. 1.63. The fitting parameters I_0 and V_0 have the values 3.65 mA and 4.68 V, respectively, for the positive polarization, and −2.6 mA and −3.12 V for the negative polarization.

The demodulating term is then found as the second-order term in the Taylor series expansion of the current around the operating point, for which $I = I_{av}$, $V = V_{av}$:

$$\Delta I = I - I_{av} = I_0 \frac{V_{RF}^2}{4V_0^2} \exp(V_{av}/V_0) \tag{1.80}$$

In (1.80) V_{RF} is the RF signal value.

A direct radio can be obtained by connecting this graphene coplanar detector to a modulated high-frequency generator and an oscilloscope, as indicated in Fig. 1.64. The radio can detect carrier wave frequencies between 100 MHz and 25 GHz at a constant power of 0 dBm (1 mW), equivalent to the emitted power of a Bluetooth standard (Class 3) radio in the range 1–5 m. In contrast to radios based on semiconductors, the direct graphene radio is able to work with a good responsivity in a huge frequency range, from the RF spectrum (starting with VHF) up to microwaves

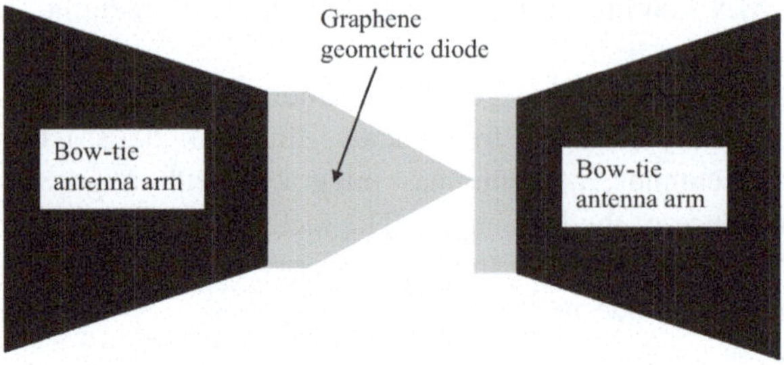

Fig. 1.61 Graphene rectenna

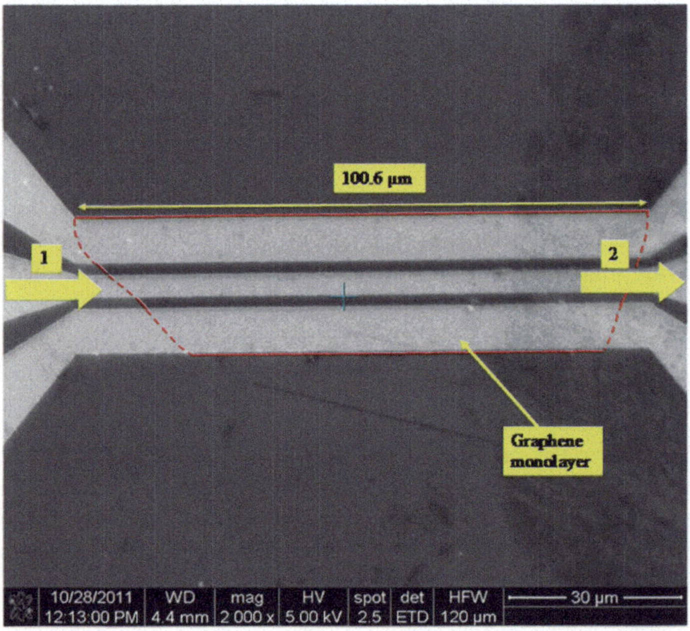

Fig. 1.62 The coplanar line graphene detector (from Dragoman et al. 2012a, b)

(L-K bands). The AM modulation is 1 kHz, which is compatible with human voice intelligible signals, and is the central frequency of audio hearing aid devices.

The dependence of the amplitude of the demodulated RF signal on frequency is illustrated in Fig. 1.65 for various DC current I_{av} values: 1 mA (gray thin line), 2 mA (black line) and 3 mA (gray thick line). If the responsivity is defined as the ratio between the detected signal and the RF incident power, the maximum responsivity of the graphene radio, of 1100 V/W, is obtained at 3.5 GHz. Moreover, the CPW graphene demodulator can detect voltages in the range (0.005 V, 0.2 V) in the 3–10 GHz spectrum in the absence of an applied DC current, suggesting that the

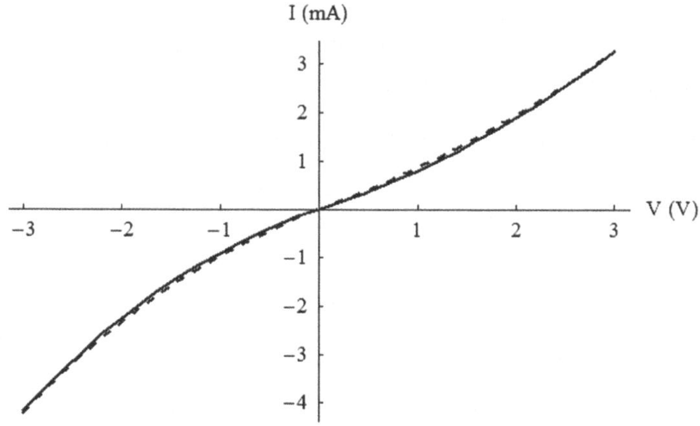

Fig. 1.63 Current-voltage dependence of the coplanar line patterned over graphene monolayer (from Dragoman et al. 2012a, b)

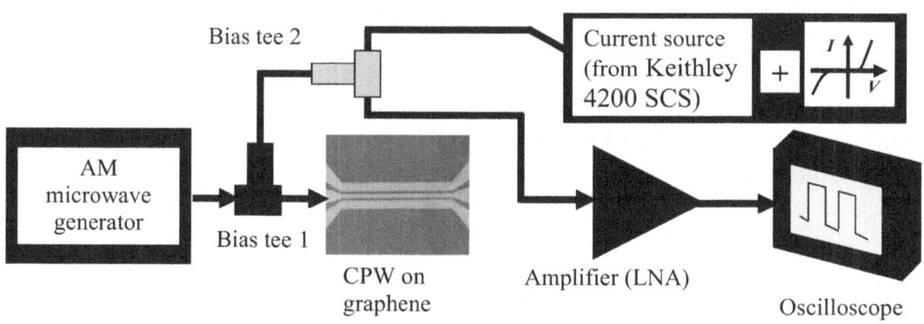

Fig. 1.64 A graphene monolayer direct radio (from Dragoman et al. 2012a, b)

Fig. 1.65 Amplitudes of detected signal using the graphene radio (from Dragoman et al. 2012a, b)

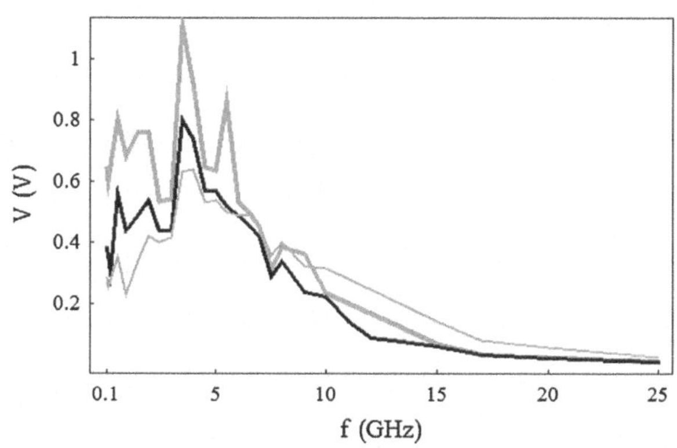

graphene coplanar line could work as rectenna in a frequency range where the majority of wireless communications, radar and power transmitters emit.

The results above show that a single sheet of matter—the graphene monolayer—is able to detect high-frequency signals with a very good sensistivity. The first graphene integrated circuit receiver with a carrier frequency of 4.3 GHz was later developed and is represented in Fig. 1.66 (Han et al. 2014). It has an area of 0.6 mm^2 and consists of a two-stage amplifier and a mixer stage. All three transistors T1, T2 and T3 are graphene transistors, with gate lengths of 900 nm and device widths of about 12 μm. The conversion gain of the circuit with 4.3 GHz RF input signal is 10 dB with a low LO input power of 2 dBm, and the gain of the amplifier is around 3 dB. Except the technological performance of integrating few graphene transistors, this receiver shows nothing special regarding its performances, which are well below the existing integrated receivers working in the same frequency band with the same bandwidth. This is a consequence of the fact that graphene transistors in the integrated circuit have poor RF performances.

The results were extended rapidly towards THz frequency range detectors using antennas integrated with graphene transistors. More exactly, THz fields are detected by shaping the gate and source electrodes of graphene FETs into various metallic antennas. The first experimental demonstration of this concept is the detection of a 0.3 THz signal at room temperature using a log-periodic circular-toothed antenna with an outer radius of 322 μm and a grounded source electrode. The channel of the transistor was rather long, of 10 μm. The detected signal at the thin drain contact can be written as (Vicarelli et al. 2012):

$$\Delta V = \frac{V_a^2}{4\sigma(V_0)} \frac{d\sigma(V_G)}{dV_G}\bigg|_{V_G=V_0} \tag{1.81a}$$

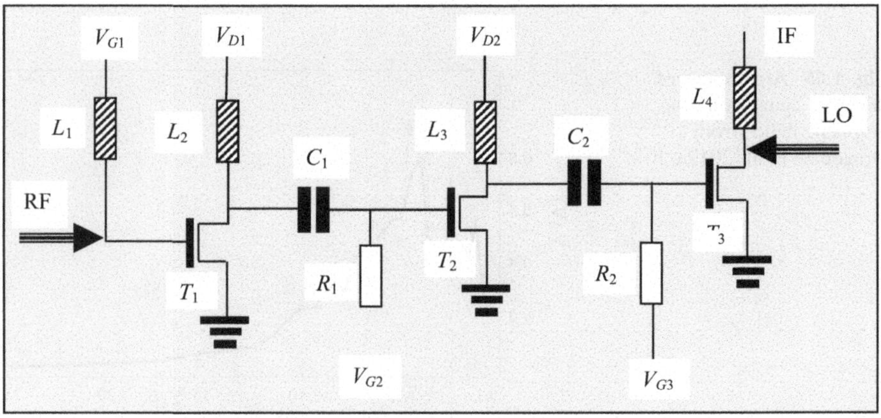

Fig. 1.66 The 4.3 GHz integrated graphene circuits

where the gate voltage is expressed as

$$V_G(x,t) = V_0 + V_a \exp(-kx)\cos(\omega t - kx) + \frac{V_a^2}{4\sigma(V_0)} \frac{d\sigma(V_G)}{dV_G}\bigg|_{V_G=V_0} [1 - \exp(-2kx)] \tag{1.81b}$$

Another type of antenna—the split bow-tie antenna—was implemented in the same way, consisting of a FET with a graphene channel, with the electrodes in the shape of a bow and capacitive coupling between gate and drain. Figure 1.67 displays the split bow-tie THz graphene detector. In this case, the channel length was shorter, having 2.5 μm, and the gate width was 2 μm. The THz signals were detected at a higher frequency, of 0.6 THz, with a noise-equivalent power (NEP) of 515 pW/ $Hz^{1/2}$ and a responsivity of 14 V/W (Zak et al. 2014).

Similarly, a logarithmic-periodic antenna with an interdigitated electrode, consisting of 4 electrode pairs deposited over a graphene flake located in the gap is able to detect THz radiation pulses despite the striking simplicity of this device. This room-temperature THz detector shows a rise time of only 50 ps (Mittendorff et al. 2013). An analogous antenna to that in (Mittendorff et al. 2013) but with bow-tie-shaped instead of logarithmic-periodic arms is represented in Fig. 1.68.

Also, very sensitive THz detection was demonstrated recently at room temperature using the photothermoelectric effect in graphene (Cai et al. 2014). This effect

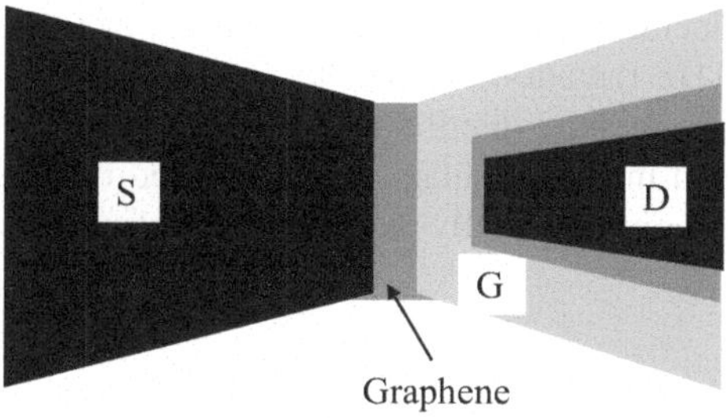

Fig. 1.67 Bow-tie antenna electrodes of a graphene FET for THz radiation

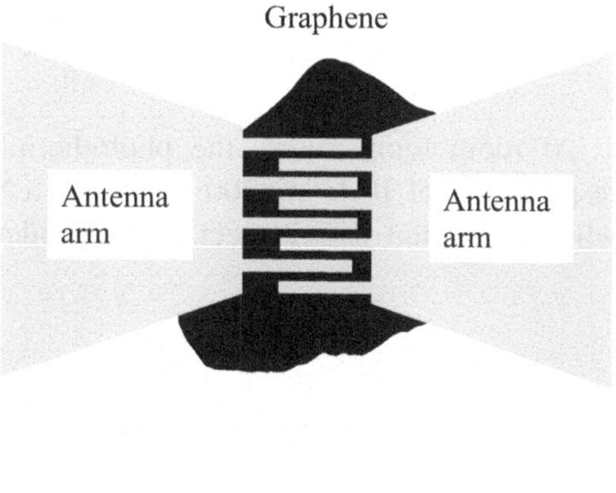

Fig. 1.68 Graphene-based THz pulse detector

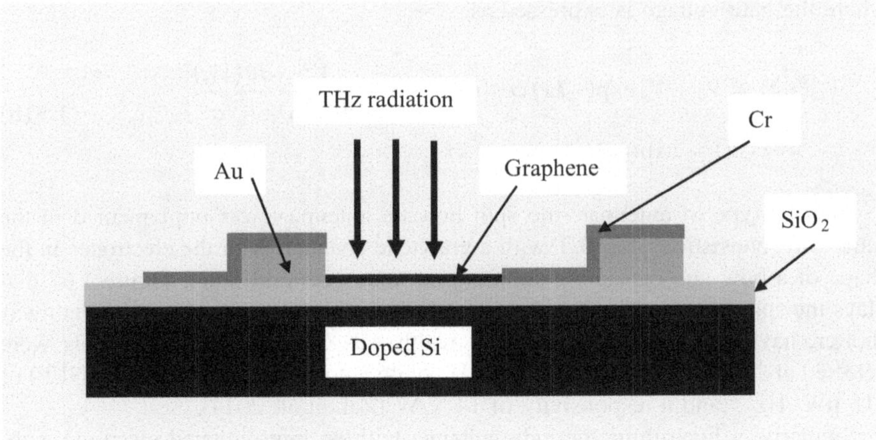

Fig. 1.69 Photothermoelectric THz detection

is based on a significant interband absorption of graphene in the THz region, as well as on a carrier heat capacity much smaller than in bulk materials, implying an increased temperature gradient for the same absorbed energy. Two dissimilar metals with a $L = 3$ μm gap between them were used in experiments to contact graphene, as shown in Fig. 1.69. These metals are in fact overlapping regions of Cr and Au, fabricated via consecutive and different angle evaporations of these metals. The operation principle is based on the fact that the carriers are heated by the incoming THz radiation, while the metallic contacts act as heat sinks, the result being a temperature distribution $T(x)$ along the device. Then, due to the dissimilar contacts and thus dissimilar contact-induced doping, the Seebeck coefficient S of graphene becomes spatially asymmetric, and a potential gradient

$$\nabla V(x) = -S\nabla T(x) \tag{1.82}$$

develops due to charge diffusion. The response/output signal of the THz detector is

$$v = \int_0^L \nabla V(x)dx \tag{1.83}$$

At room temperature, the photothermoelectric graphene THz detector has a responsivity of 10 V/W (700 V/W) and a NEP of 1100 pW $Hz^{-1/2}$ (20 pW $Hz^{-1/2}$) when calculated with respect to the incident (absorbed) power.

1.3 Electronic Sensors on 2D Carbon-Based Materials

Graphene is used in many electronic sensors due its exceptional physical and chemical properties, which are listed below (Liu et al. 2015a):

- high and tunable electrical conductivity
- high electron transfer rate due to edges and defects
- mechanical flexibility and strength (two orders of magnitude stronger than steel)
- high surface area (2630 m^2/g)
- chemical stability
- absence of metallic impurities like in carbon nanotubes and other nanomaterials
- large electrochemical potential window (2.5 V in 0.1 mM phosphate buffer saline solution)
- high thermal conductivity
- fast and strong response in sensing applications

All these properties make graphene a good candidate for many sensing applications. However, the graphene surface needs to be functionalized in order to respond to the many requirements of sensing applications.

By smart processing of its surface, such as oxidation and heteroatomic doping, graphene is enriched with functional materials on its surface and edges. Covalent, noncovalent and other functionalization methods are used in this respect. In principle, each sensor requires a certain functionalization, but some functionalization methods are more frequently used than others and are very successful, and therefore will be presented briefly before describing the main electronic sensors on graphene. The optical sensors based on graphene will be presented in the next section.

Covalent functionalization destroys the conjugated sp^2 structure of graphene and could even immobilize the delocalized π electrons, thus suppressing the graphene conductivity. For example, highly reactive free radicals to be used for addition reactions, which fix aryl-addends on graphene surface, are produced by diazonium salts. These reactions transform the sp^2 structure of graphene in sp^3 and the functionalized graphene becomes and insulating/nonconducting material (Bekyarova et al. 2009). A typical covalent functionalization involves the attachment of carboxyl groups on graphene surfaces, which react in turn with amino groups of a target molecule (DNA, enzyme or proteins). This is the functionalization used for the majority of biosensors that detect electronically DNA or proteins via graphene FETs, for example. Also, graphene oxide is functionalized with thionyl chloride for reaction with hydroxyl or amino groups, whereas nitrogen and boron are used for graphene doping.

Noncovalent functionalizaton, on the other hand, preserves the original sp^2 structure of graphene, and hence its electrical properties. The main procedure is based on hydrophilic and hydrophobic interactions, and π-π stacking. The review (Georgakilas et al. 2012) contains more details on the chemical reactions involved in graphene functionalization and many useful references.

Graphene doping is interesting for a series of applications. For example, a large Fermi level shift, of up to 0.9 eV, corresponding to a high doping concentration, is obtained using ion gels (Chen et al. 2011b) such as 1-ethyl-3-methylimidazolium bis(trifluoromethylsulphonyl)i-mide ([EMIM][TFSI]) as gate dielectrics (see Fig. 1.70).

One of the main reasons of graphene doping, except carrier modulation in transistors and diodes and other tunable devices, is to reduce the surface resistance of graphene. For instance, doped graphene could replace ITO (indium tin oxide) as transparent electrode for solar cells, which would be beneficial because indium is a scarce raw material used in tactile displays, i-phones, laptops and many solar cells. An example of chemical doping of graphene is the use of bis bis (trifluoromethanesulphonyl)-amide [($(CF_3SO_2)_2NH$)] (TFSA) to increase the power conversion efficiency of a monolayer graphene/Si solar cell from 1.9 to 8.6% (Miao et al. 2012) due to surface resistance reduction as well as Schottky barrier/built-in field enhancement. The doping of graphene is performed by spin-casting TFSA (20 mM in nitromethane) for 1 min at 1000–1500 rpm. The Fermi energy level is shifted by 0.7 eV due to TFSA doping, the series resistance decreases with 30% and the built-in bias increases from 0.36 to 0.56 V. The sheet resistance of five-layer graphene doped with TFSA on a flexible (polyethylene terephthalate) PET substrate is reduced even more significantly, from 240 to 90 $\Omega\ sq^{-1}$ (Kim et al. 2013a).

Because graphene is highly transparent, tactile displays and even stretchable i-phones are among its first applications. The tactile display is in fact an actuator composed from graphene electrodes and a dielectric elastomer substrate. When DC voltages are applied on graphene electrodes, the elastomer is bumping up producing actuation, and it preserves its physical properties up to 25% stretching (see Fig. 1.71).

A cost-effective solution for transparent electrodes used for touch screens is to use silver nanowires coated with graphene, which reduces the amount of required nanowires by 50 times compared to a high-density nanowire network. Thus, the electrical performance of silver nanowires, with ultra-low densities, is enhanced by graphene coating. The Langmuir-Schaefer technique is used to wrap graphene around silver nanowires. As a result, the sheet resistance, of around 600 kΩ/sq of a low-density Ag nanowire network having a transmittance of 97%, is drastically

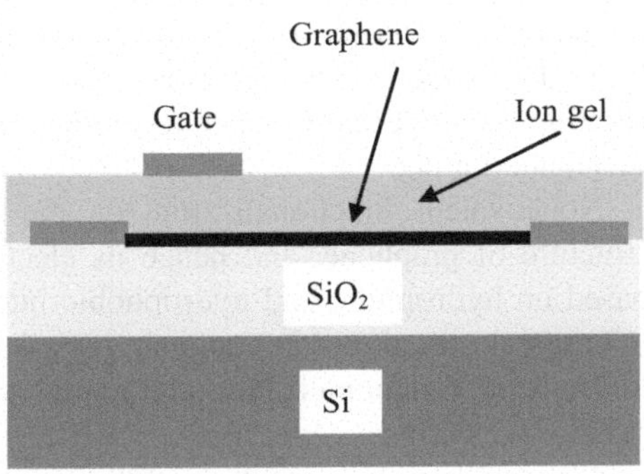

Fig. 1.70 Graphene doping with ion gel

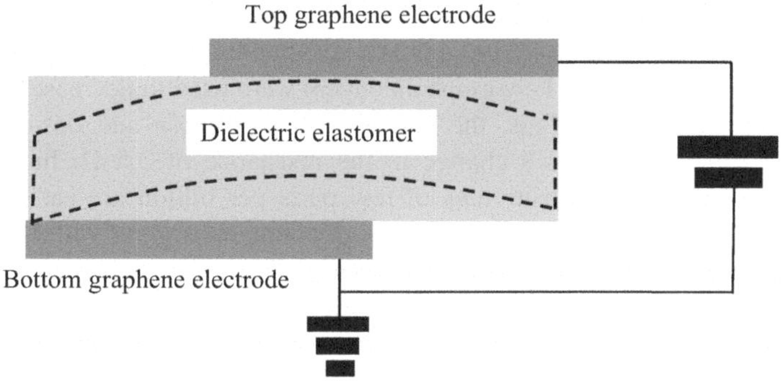

Fig. 1.71 Graphene actuator for tactile display

reduced by graphene coating, reaching up to 200 Ω/sq, without affecting too much the transmittance, which remains at a high level, of 93%. Similarly, high-density silver nanowires having initially a 40 Ω/sq sheet resistance and a transmittance of 93%, attain a sheet resistance of 28.7 Ω/sq and a transmittance of 87% when covered with graphene. These results are explained by the fact that the electrical conduction of the hybrid structure is increased due to percolation doping. Finally, the touch screen is deposited by laser ablation (Jurewicz et al. 2014).

Since the specific surface area of graphene is 2630 m^2/g—the highest of all materials, graphene is a good material for sensing applications, especially as gas sensors or biosensors (Hill et al. 2011; Yavari and Koratkar 2012; Dragoman and Dragoman 2012). In particular, graphene gas sensors are sensitive enough to detect even a single molecule of gas (Schedin et al. 2007). Gas detection is based on specific changes in graphene resistivity for different gases, which after adsorption on graphene in an interval of typically about 200 s induce a variation of resistivity with +4% for NH_3, −4% for NO_2, or −1% for H_2O at a gas concentration of 1 ppm. A common graphene gas sensor configuration has the shape of a Hall bar, as illustrated in Fig. 1.72.

In this device, Hall measurements have demonstrated that some gases are graphene donors (NH_3, CO, and ethanol), i.e. in these cases $\Delta\rho/\rho > 0$, while others,

Fig. 1.72 Graphene Hall constriction (fabricated by oxygen plasma etching) acting as a gas sensor

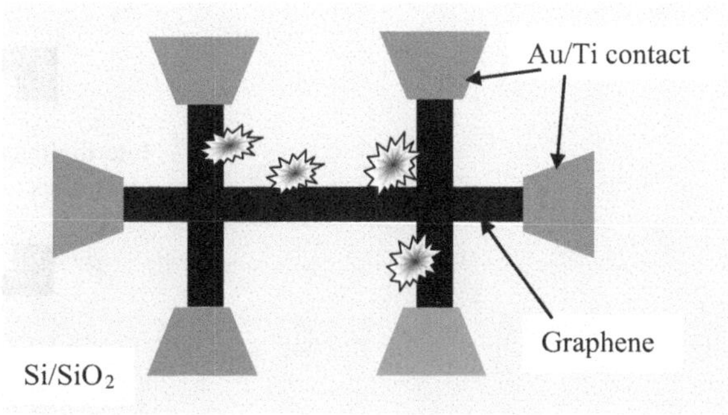

for which $\Delta\rho/\rho < 0$, are acceptors (NO_2, H_2O vapors and iodine). The detection limit of the graphene gas sensor is 1 ppb, corresponding to a variation of graphene resistivity of $\Delta\rho/\rho \propto 10^{-4}$. Few graphene layers were used to decrease the contact resistance around 100 Ω. Thus, the variation in adsorption and desorption of a single NO_2 molecule shows a change in the resistance of 2.5 Ω. In reality, for normal electronic noses, sensitivities of few parts per billion are easily obtained with a simple sensor geometry made from a graphene monolayer with two metallic contacts or interdigitated electrodes, as shown in Fig. 1.73.

More recently, researches have found a surprising result, that graphene defects, such as an isolated grain boundary, have 300 times higher sensitivity for detecting adsorbed gas molecules than a single-crystalline graphene grain. Graphene and the grain boundary are grown via CVD at normal atmospheric pressure, and controlled growth time. Under such conditions, single-crystalline graphene grains with a generally hexagonal shape grow individually and eventually merge into an individual grain boundary. Much higher gas detection sensitivity is achieved due to accumulation of adsorbed gas molecules at the boundary between graphene grains, favored by the appearance of a stress-relaxation-induced corrugation in this region, which is equivalent to local doping/electric field gradient (Yasaei et al. 2014).

Several grain boundaries connected in series can be obtained and further connected to obtain a very sensitive gas sensor, i.e. an electronic nose. Such a gas sensor containing two grain boundaries is illustrated in Fig. 1.74. The sensitivity of the sensor is defined as $S = (R - R_0)/R_0$, with R and R_0 the resistance of the device after and before gas exposure, respectively. Taking into account in a more detailed manner the different contributions to the total resistance, the sensitivity of the electronic nose can be written as

$$S = \frac{N\Delta R_{GB} + 2N\Delta R_{G1} + (N-1)\Delta R_{G2}}{NR_{GB} + 2NR_{G1} + (N-1)R_{G2}} \tag{1.84}$$

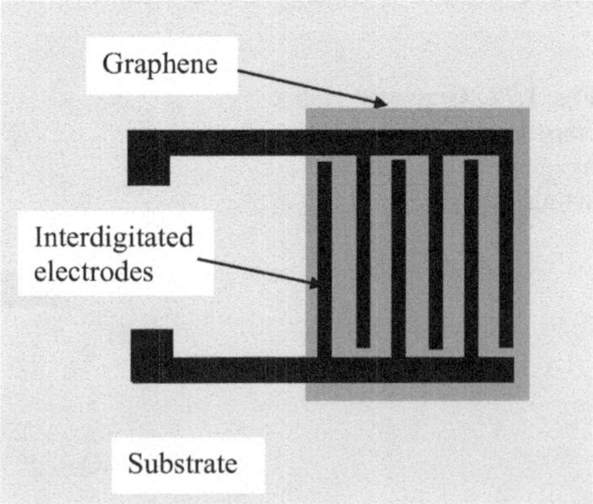

Fig. 1.73 An interdigitated electrode for graphene gas sensing

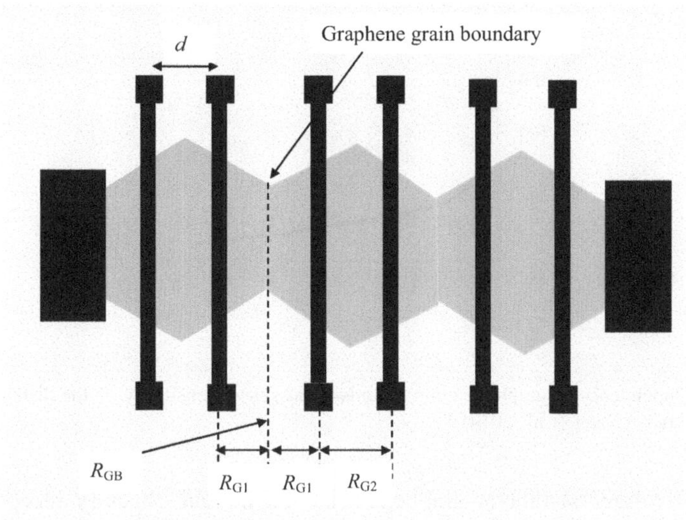

Fig. 1.74 Gas sensor containing two graphene grain boundaries

where N is the number of grain boundaries, R_{GB} is the resistance of an isolated grain boundary, and R_{G1} and R_{G2} are the resistances of the trapezoidal graphene region and of the single-crystalline region, respectively. From (1.84) it follows that when $N = 1$ the terms containing R_{G2} vanish and the sensitivity increases significantly. The dependence of sensitivity on the number of grain boundaries is displayed in Fig. 1.75. Also, Fig. 1.76 indicates that the distance between electrodes, d, is critical for obtaining a high sensitivity.

As an example of an ultrasensitive electronic nose, a bilayer graphene transistor was fabricated on a flexible PET substrate (Park et al. 2012). The transitor has a ionic liquid gate consisting of a phosphate-buffered solution (PBS) with a pH of 7.4, the sensing layer being obtained by integrating the modified (by oxygen and

Fig. 1.75 Sensitivity dependence on N of a graphene grain boundary gas sensor (after Yasaei et al. 2014)

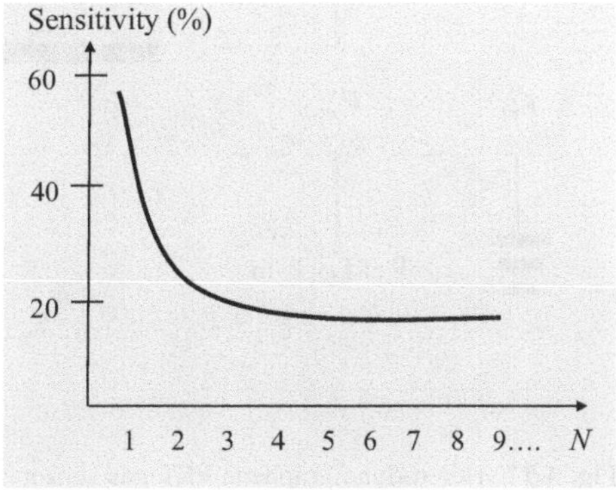

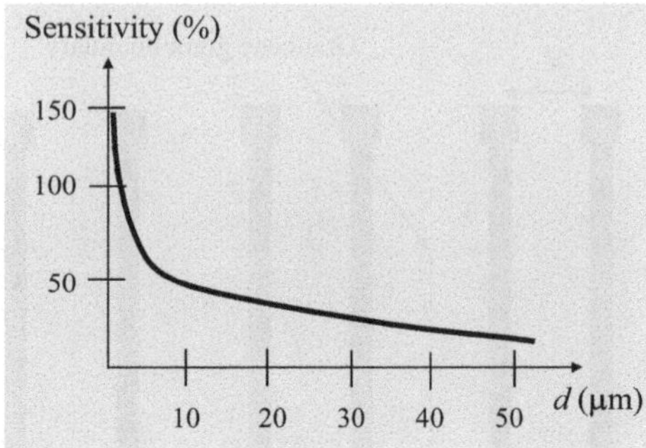

Fig. 1.76 Dependence of graphene grain boundary gas sensor sensitivity on the distance between electrodes (after Yasaei et al. 2014)

ammonia treatments) bilayer graphene with the human olfactory receptor 2AG1 (hOR2AG1:OR). OR binds selectively to the amyl butyrate (AB) odorant. This flexible nose is so sensitive that it is able to detect an odorant with atom resolution, with a minimum detection limit of 0.04 fM. The sensing is based on the change of the transistor drain conductance with the AB binding quantity.

Very sensitive and very rapid vapor sensors are obtained using heterodyne graphene sensors (Kulkarni et al. 2014). In this case, an AC signal V_ω, with frequency ω, is injected in the source of a graphene FET with DC grounded source and gate, as shown in Fig. 1.77. The electrostatic coupling between the gate

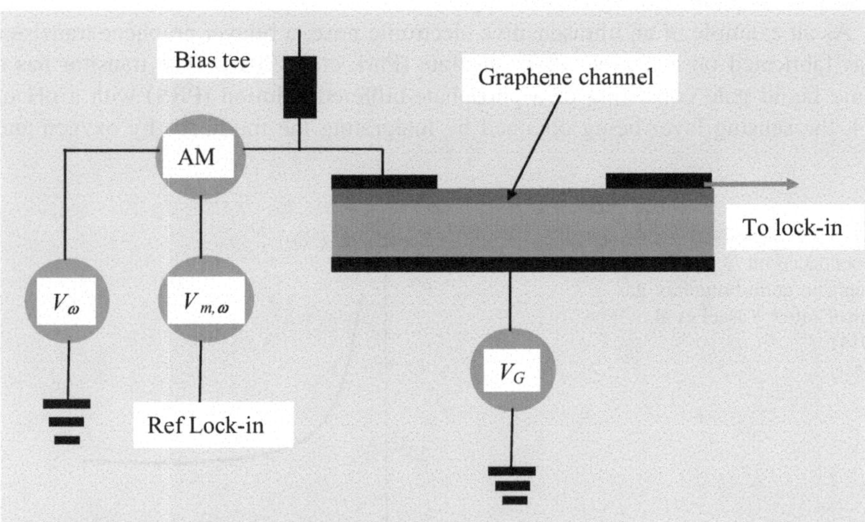

Fig. 1.77 Heterodyne graphene FET gas sensor

electrode and the graphene channel generates a gate-induced modulation of the charge density given by

$$Q_{G,\omega} = C_G V_{G,\omega} = -C_G V_\omega / 2 \tag{1.85}$$

where C_G is the gate capacitance. When a molecule is adsorbed, dipole-induced charge densities $Q_{m,\omega}$ appear, modulated at the frequency ω. The frequency mixing between V_ω and the total modulated charge is expressed by the mixing current

$$I_{mix} = G_\omega V_\omega = (\mu W/L)(-C_G V_\omega/2 + Q_{m,\omega}) V_\omega \tag{1.86}$$

For vapor sensing, the relevant quantity is the change in the mixing current, given by

$$\Delta I_{mix} = (\mu W/L) Q_{m,\omega} V_\omega \tag{1.87}$$

To improve the noise performance, an AM modulation can be introduced. The AC signal frequency is 100 kHz and the AM modulation frequency is of 1.43 kHz in the device studied in (Kulkarni et al. 2014). This sensor was tested for dichloromethane, showing a response time of 0.61 s, for ethanol, with a response time of 0.92 s, and for other vapors with similar response times. The sensor is able to detect airborne chemicals, either exhaled or released through skin, and various dangerous gases.

There are also mechanical sensors based on graphene, which detect parameters such as pressure and strain. Pressure sensors used as touch control displays were presented earlier. However, for a gentle touch and small objects' manipulation, pressure in the range of 10–100 kPa need to be detected. Laser-scribed graphene (LSG), which are loosely stacked graphene layers with a foam-like structure, can be used as sensing material for these purposes (Tian et al. 2015). The sensor is displayed in Fig. 1.78. The sensitivity of the LSG graphene sensor is defined in terms of the relative variation of the conductance as

$$S = \delta(\Delta G/G_0)/\delta P \tag{1.88}$$

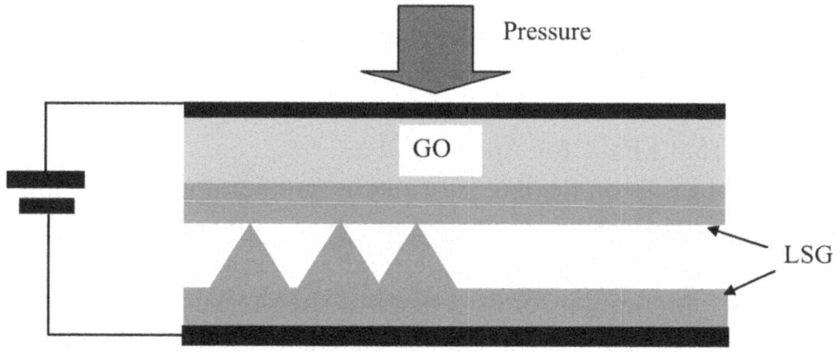

Fig. 1.78 LSG graphene pressure sensor

its value being 0.96 kPa^{-1} for pressures P lower than 50 kPa, while the response and releasing times at 26 kPa are of 72 and 0.4 ms, respectively.

Using the same LSG technology, graphene earphones on PET were fabricated for communications for both human and animals (Tian et al. 2014). Humans can hear sounds in the bandwidth 20 Hz–20 kHz, whereas animals communicate in the ultrasound spectrum: bats up to 160 kHz, whales at 160 kHz, while dolphins generate sounds at 300 kHz. The graphene earphones have a frequency response in the range 100 Hz–50 kHz and low fluctuations (±10 dB), whereas graphene sound emitters could reach 1 MHz.

Arrays of pressure/temperature sensors, called electronic skin in analogy with human skin, have a wide range of applications in bionics, robotics and high-tech industries. An e-skin made from reduced graphene oxide foam (RGOF) obtained from a frozen GO solution reduced with hydriodic acid (HI) followed by washing and drying was recently reported (Hou et al. 2014). RGOFs are porous due to the ice template, and the networks preserve their patterns after ice melting and removal. RGOFs are freely standing, flexible, and have a sheet resistance of 1.5 Ω/sq, this remarkable value being due to the fact that HI reduction induces iodine doping. RGOFs can be compressed up to 65% by a strain of 2 kPa and recover rapidly their initial shape and physical properties; for example, the electrical properties are recovered after 0.05 s. As such, the RGOF can be tuned by human touch (1–10 kPa). Moreover, RGOFs response discriminates between human touch and otherwise applied pressure via surface temperature difference between humans and inanimate objects, which indicates a temperature-activated sensing mechanism.

As electronic skin, the RGOF is contacted on the upper and lower surfaces and placed on PET, at a temperature of 20 °C, as schematically represented in Fig. 1.79. The upper levels of this artificial skin are heated at different temperatures. Rather high temperature differences can be applied between the heated surface, along x, and other areas of the devices (along the y and z directions), inducing corresponding current flows due to the high thermoelectric effect in graphene and its network formed in RGOF. For example, if a temperature difference of about 30 °C is applied between electrodes along the z direction, a current of 9 μA is measured between the corresponding electrodes (a comparable current is measured between electrodes along the y direction), if the heating surface of RGOF sample is 5.6 cm^2. Based on this principle, RGOF is able to discriminate between human body and other objects, such as metals, plastic or glass, when the applied pressure alone is not high enough to induce significant changes in the resistance. The sensor responds in this case only to human body touch, because the body temperature is higher than the surroundings (Hou et al. 2014). The sensitivity of this device to pressure, defined as $\delta(\Delta I/I_0)/\delta\Delta p$, is 15.2 kPa^{-1} for Δp = 300 Pa.

Placing arrays of FETs on flexible substrate is another approach to implement an e-skin (Sun et al. 2014). Such a device involves only graphene and a ion gel dielectric, and consists of a top cover layer and a backplane, separated by 100 μm thick epoxy; this configuration forms a pixel of the pressure sensor matrix, which contains 4 × 4 pixels. The top cover contains a square graphene region, the contour of which is represented by a dashed line in Fig. 1.80, placed on a flexible PET

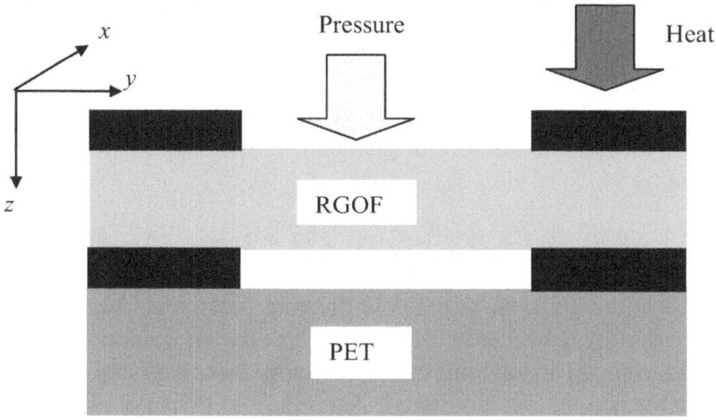

Fig. 1.79 Electronic skin based on RGOF

substrate. The backplane contains a graphene FET, which acts as pressure sensor. The graphene FET is patterned in a coplanar-gate configuration, and consists of a single graphene strip on which an ion gel is deposited. The graphene region below the ion gel is the FET channel, the remaining graphene strip regions acting as drain and zigzag-shaped source electrodes. The drain is connected to the bit line and the gate electrode, which is also a graphene strip, is connected to the word line of the device (see Fig. 1.80). The gate dielectric is in this case an ion gel formed from poly (ethylene glycol) diacrylate (PEGDA) monomers, the photo-initiator 2-hydroxy-2-methylpropiophenone (HOMPP) and the ionic liquid 1-ethyl-3-methylimidazolium bis(trifluoromethyl sulfonyl)imide ([EMIM][TFSI]) with a weight ratio 7:3:90. When pressure is applied, the square graphene on the top cover is in contact with the bottom zigzag source electrode, the resistance between source and drain is reduced, and thus the conductance increases. The transistors work at 2 V and have a pressure

Fig. 1.80 One pixel of a graphene-based e-skin

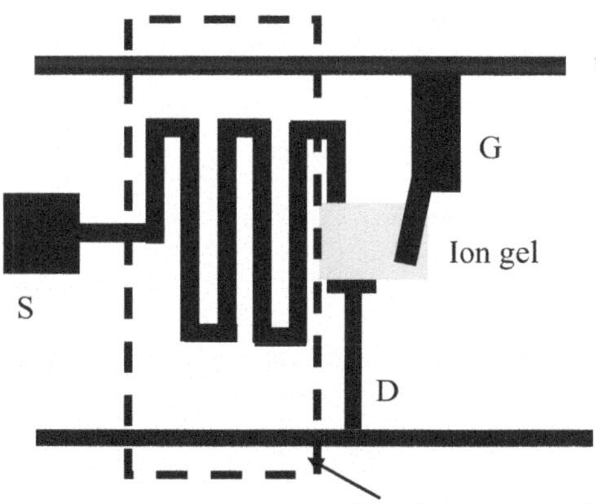

sensitivity of 0.12 kPa^{-1}. The drain conductance is about 5 μS at 10 kPa, 8 μS at 20 kPa, and 14 μS at 30 kPa.

Because there is a close relation between pressure and strain, graphene is a good candidate for strain sensors also (Zhao et al. 2013). The relation between the variation of the electrical resistance of graphene and the applied mechanical strain ε can be written as

$$\Delta R/R = (1+2\nu)\varepsilon + \Delta\rho/\rho = (1+2\nu)\varepsilon - 2\Delta v_F/v_F \quad (1.89)$$

where $\nu = 0.2$ is the Poisson ratio, ρ is the resistivity and $\Delta\rho$ and Δv_F are the changes in resistivity and Fermi velocity, respectively, caused by strain. Strain sensors are widespread in buildings, bridges, airplanes, cars and anywhere where mechanical deformations are monitored. According to (Zhao et al. 2013) the strain market is exceeding 5 billions dollars. A typical strain sensor configuration based on suspended graphene is represented in Fig. 1.81.

The gauge factor describing the sensitivity of strain sensors is defined as the ratio between the electrical resistance variation and the mechanical strain:

$$G = (\Delta R/R)/\varepsilon \quad (1.90)$$

We have pointed out already that graphene has exceptional mechanical properties, the third-order elastic stiffness of this material being −2 TPa, with an intrinsic strength of 130 GPa at 25% in-plane tensile strain (Lee et al. 2008). A suspended graphene membrane used as a strain sensor shows a 5% change in the resistance ($\Delta R/R$) for 3% strain, corresponding to a gauge factor of 1.9 (Huang et al. 2011b). For CVD grown graphene transferred on PDMS substrates, it was shown that the electrical conductance behaves reversibly as the elastic tensile strain varies up to 4.5% (Fu et al. 2011) with a high gauge factor of 150.

The piezoresistivity of graphene is used in many strain sensor types. For example, large graphene areas in which graphene growth is not completed, i.e. consisting of graphene sheets connected together in a network, are sensitive strain detectors. When compressive or tensile strain is applied to the graphene area, the overlapping areas between graphene sheets change, inducing a change in resistance.

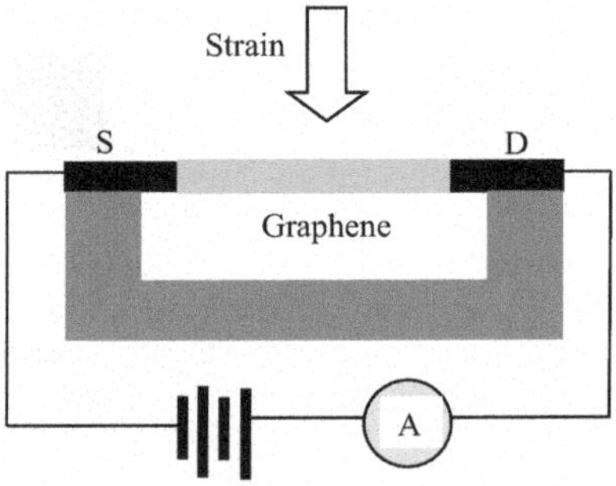

Fig. 1.81 Graphene strain sensor

Tunneling between adjacent graphene sheets is another mechanism for obtaining an exponential dependence of the electrical resistance as a function of applied strain.

A piezoresistive sensor containing a SiNx membrane on which a graphene meander resistance is patterned is depicted in Fig. 1.82 (Zhu et al. 2013a). The pressure sensing mechanism is based on inducing strain in graphene by applying pressure across the membrane (see Fig. 1.83). The maximum tensile strain, of about 0.25%, appears at the middle of each edge of the membrane. The thickness of the square SiNx membrane is of 100 nm, its length is 280 μm with a residue stress of 50 MPa, a Young modulus of 200 GPa and a Possion ratio of 0.22.

Under an applied strain, the resistance R of the meandering graphene resistor, measured with a Wheatstone bridge, changes with ΔR. The parameter that is measured is the voltage drop across the graphene resistor for an input voltage of V_{in} = 2.5 V. The output voltage is given by

$$V_{out} = V_{in}\Delta R/(2R + \Delta R) \tag{1.91}$$

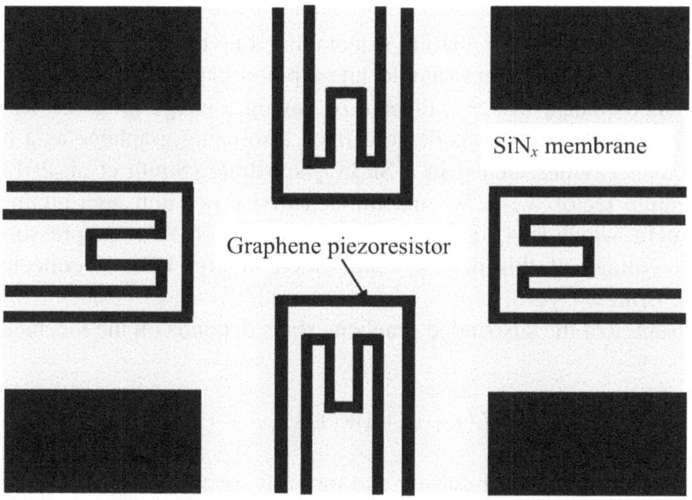

Fig. 1.82 Pressure sensor containing four graphene piezoresistors

Fig. 1.83 Principle of the graphene pressure sensor

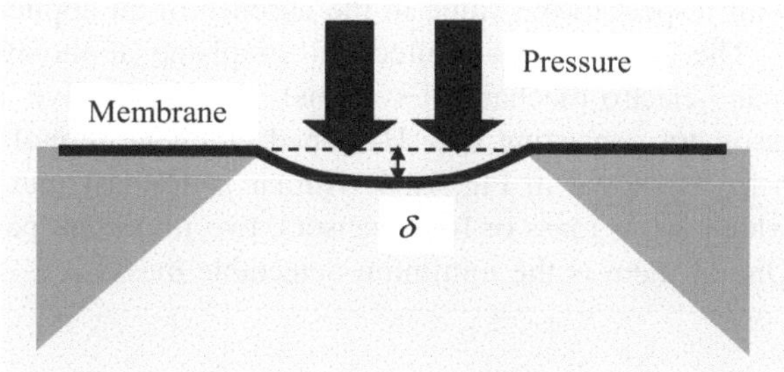

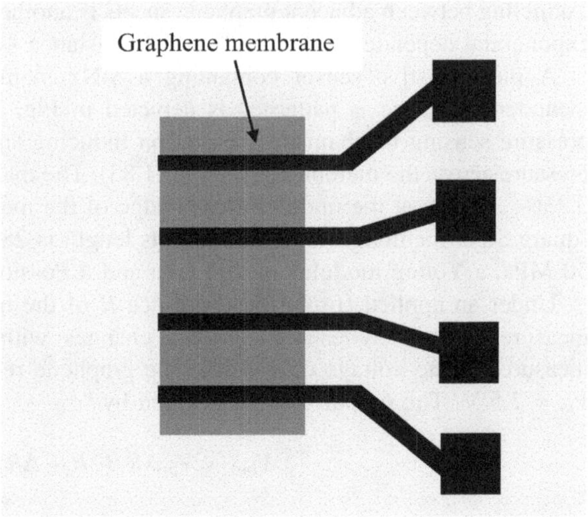

Fig. 1.84 Pressure sensor on a graphene membrane

The linearity of the above relation is maintained up to 500 mbar (50 kPa) with a sensitivity of 8.5 mV/bar. For example, an output voltage of 5 mV is obtained for a 0.4% resistance change in 0.9 s, the sensor having a gauge factor of 1.6.

The piezoresistive effect was demonstrated also using graphene as a membrane suspended over cavities etched in a Si/SiO_2 substrate (Smith et al. 2013). In this case, the gauge factor was 2.92 and the sensitivity per unit area attained almost 4 μV/V/mmHg, which is 20–100 times better than in conventional pressure sensors. The output voltage of this device, represented in Fig. 1.84, is collected with a Wheatstone bridge.

The resistance of the suspended graphene sheet depends on the mechanical strain as

$$R_{g,suspended} = L(1 + \varepsilon_{xx})/[W(1 + \varepsilon_{yy}) \times (2eN(\varepsilon)\mu(E))] \quad (1.92)$$

where N and μ are the carrier density and mobility, respectively, and ε_{xx} and ε_{yy} are the strain tensor's components parallel and perpendicular to the carrier transport direction. The induced strain is uniaxial because $\varepsilon_{xx} \gg \varepsilon_{yy}$. The value of the resistance in Smith et al. (2013), of about 0.2 kΩ at 1000 mbar, changes with 0.6% with respect to the value in the absence of an applied pressure.

The piezoresistive effect in graphene monolayers is exploited in NEMS (nano-electro-mechanical-systems) as an effective readout method, the NEMS resonator consisting of a H-shaped graphene monolayer clamped at the four corners, as shown in Fig. 1.85 (Kumar and Bhaskaran 2015). In NEMS resonators, which act as mass or force sensors, two important parameters must be determined. One of them is the minimum detectable mass:

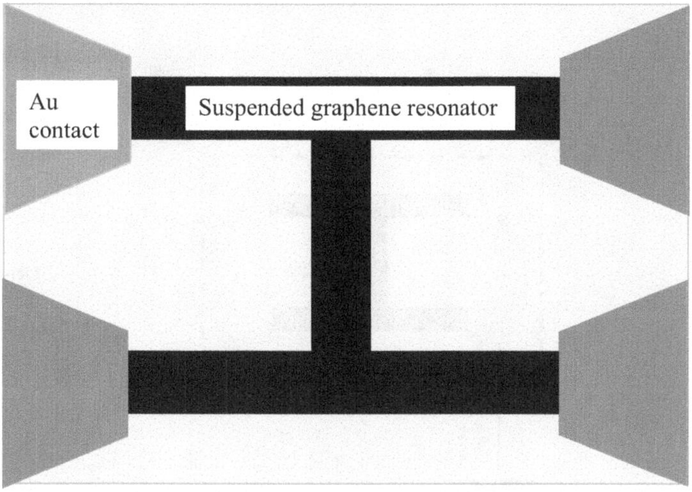

Fig. 1.85 The graphene monolayer suspended NEMS resonator

$$\delta m = 2m_{eff}\sqrt{\Delta f/Q \times \omega_r}10^{-(DR/20)} \quad (1.93)$$

where m_{eff} is the mass of the resonator, Q is the quality factor, ω_r the resonance frequency, Δf is the bandwidth, and DR the dynamic range. The other parameter is the minimum detectable force, defined as

$$F_{\min} = \sqrt{4k_BTk/\omega Q} \quad (1.94)$$

where $k = m_{eff}\omega^2$ is the spring constant and T the temperature.

The mechanical resonator with a resonant frequency of about 1.2 MHz has a total length of 1.8 μm and a total width of 1.2 μm, each leg being 120–150 nm wide and 0.5–0.7 μm long. The quality factor at room temperature and 4 μTorr is around 1000. The resonator is mounted on a piezo disk and is mechanically activated by an external DC voltage (see Fig. 1.86), while a bias voltage is applied across the two legs on the same side of the resonator to transduce the change in resistance of the resonator due to mechanical action into voltage. The minimum detectable mass is 1.4 zg (10^{-21} g), and the minimum detectable force of this mechanical sensor is 16 aN/(Hz)$^{1/2}$ for a spring constant of 10^{-4} N/m.

Wearable electronics is one of the most advanced nanotechnologies focused on new devices that combine mechanical and electronic properties of nanomaterials mounted on skin or near the skin. Nanomaterials for such applications must have special properties. In particular, they should be stretchable, bendable, and deformable at any motion. Such devices, integrated into clothing or directly on skin, have applications in monitoring of finger and muscle motions, sensing the posture,

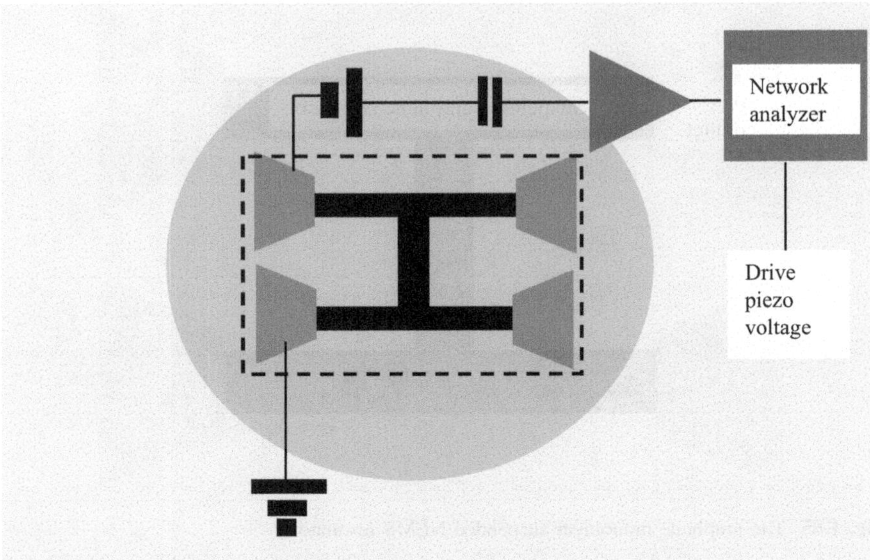

Fig. 1.86 Measurement set-up of the graphene NEMS resonator

movement and breathing. Medical applications range from emergency medicine up to remote monitoring of patients.

Wearable sensors must satisfy some prerequisites (Boland et al. 2014):

(i) light weight and high degree of miniaturization
(ii) high sensitivity
(iii) ability to work under high strain
(iv) fast time response

Recently, it was demonstrated that graphene-rubber composites satisfy the conditions above (Boland et al. 2014), graphene being an ideal filler for strain sensors. Graphene infiltrates commercial elastic bands, conferring them electrical properties, sensing being based in most cases on the change of the resistance with strain. The concentration or volume fraction of graphene at depth x in composites

$$\phi(x) \propto erfc(x/\sqrt{4Dt_s}) \tag{1.95}$$

is determined by the diffusion coefficient of graphene into rubber, $D = 4 \times 10^{-13}$ m^2/s, and the soak time t_s. At low graphene concentration, the resistance of the composite is 4 MΩ, but decreases at 57 kΩ at a graphene concentration of 0.56% with a soak time of 48 h. The conductivity of the composite is described by the percolation theory, where the conductivity is given by

$$\sigma = \sigma_0(\phi - \phi_c)^t \quad (1.96)$$

with t the percolation exponent, σ_0 the conductivity of the filler in a thin film configuration, and $\phi_c = 0.1\%$ the experimentally determined graphene concentration at which percolation occurs.

The resistance of the graphene rubber composite changes with orders of magnitude at strains up to 600% and beyond (see Fig. 1.87). Note that the rubber is very extensible, showing breaking at a strain of 1100%. The gauge factor is dependent also on the volume fraction, reaching 20 at 0.1%.

The above composite was used to monitor the movement of the first knuckle of index finger. The strain motion reached $\varepsilon = 40\%$ when the finger was bent few times. Breathing was also monitored by observing periodic, repetitive changes in the resistance of the composite from 1 to 2 MΩ. Also, the muscle movements were detected by wrapping the composite band around the forearm.

There are also transparent and flexible strain sensors (Bae et al. 2013), used for monitoring the motion of body parts. In this case, the strain sensor is fabricated from CVD grown graphene on Ni followed by reactive ion etching and stamping techniques, graphene being finally transferred on flexible or stretchable substrates, such as PDMS. The PDMS transmittance is 70% in the visible region. The flexible strain sensor is based on the piezoresistive effect of graphene, i.e. on the variation of the electrical resistance with strain. In this case, the resistance has a nonlinear variation with the strain, starting from 500 kΩ for 0–0.5% strain and reaching 900 kΩ for 7% strain. A rosette-like configuration containing three transparent strain sensors, each of them oriented at 120° with respect to the others, was used to detect the three independent components of the strain tensor (see Fig. 1.88). A rosette-like strain sensor array was introduced in gloves, targeting wearable electronics applications to detect finger motions.

Very effective strain sensors for wearable electronics, displaying high stretchability of up to 150%, are made in a simple way, by using different stretchable yarns on which poly(vinyl alcohol) (PVA) and graphene nanoplatelets (GNP) are deposited using layer by layer techniques (Park et al. 2015). Depending on the yarn structure, these strain sensors are able to detect large-scale body motion, such as

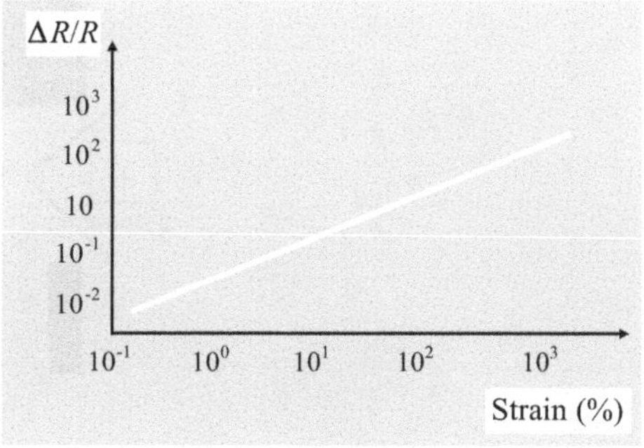

Fig. 1.87 Graphene composite change in resistance versus strain (after Boland et al. 2014)

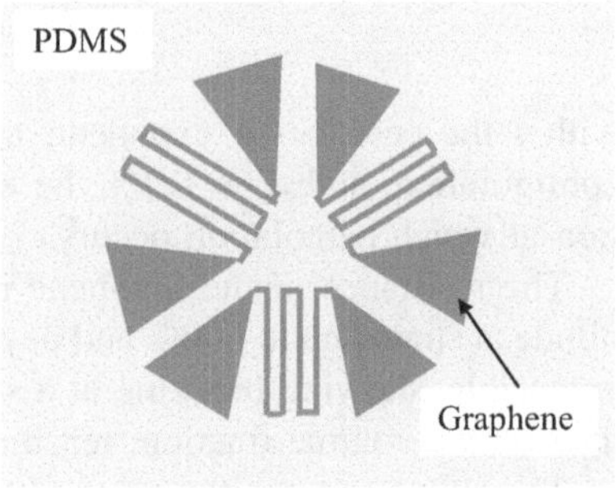

Fig. 1.88 Rosette-like transparent graphene strain sensor

bending movements of the hands, arms and legs, and small-scale body motions, such as chest and neck motions associated with breathing, speaking, and swallowing. The uncoated strain sensor based on depositing several layers of PVA and GNP on the yarn is displayed in the upper part of Fig. 1.89, whereas the sensor with a protective PDMS coating is represented in the lower part of the same figure.

Different stretchable yarns were tested in sensors coated with PDMS: (i) wool yarns (WY), (ii) rubber (RY), and (iii) nylon covered rubber (NCRY), all having good stretchability. As the strain changes between 0 and 100%, $\Delta R/R$ varies by 2000% in the case of RY and by 200% in the case of NCRY, whereas $\Delta R/R$ decreases about 10% when the strain on WY is varied between 0 and 50%. The difference in resistance increases in all cases with the number of PVA and GNP layers. In particular, small-scale motion of the throat and large-scale motion of fingers were detected by monitoring $\Delta R/R$ in time.

For wearable electronics, autonomous and miniaturized DC sources are necessary. Supercapacitances based on graphene, which are shape-tailorable, were recently reported (Xie et al. 2015). Graphene is an ideal candidate for supercapacitors since its specific surface area is 2675 m^2/g and has a very high specific capacitance (550 F/g). The supercapacitor consists of Ni nanocone arrays covered with reduced graphene oxide (RGO) sandwiched between two flexible electrodes.

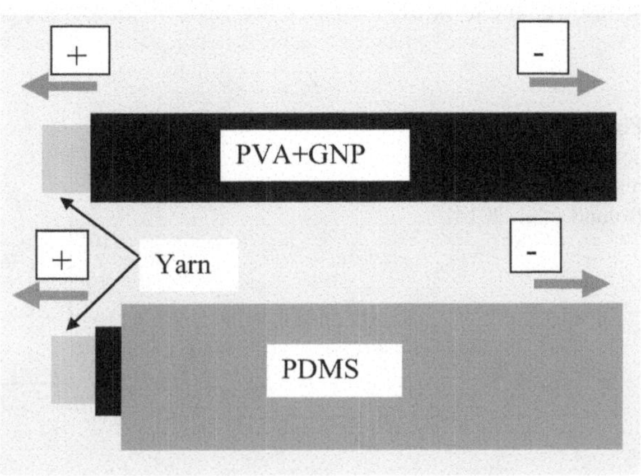

Fig. 1.89 Graphene yarn stain sensor

The last category of graphene sensors discussed in this section is the biosensor. Biosensors are well described in our book about Bionanelectronics (Dragoman and Dragoman 2012), reprinted in China at 2015. However, few years ago graphene biosensors were in infancy. Therefore, this topic is discussed in more details here.

A biosensor is a device formed from two parts: (i) a bioreceptor, which is the sensing part, connected to (ii) the detector, which transforms the information about the sensed analyte into a signal (optical, electrical, etc.). The configuration of the biosensor is represented in Fig. 1.90.

Depending on the type of bioreceptors, we have immunobiosensors, geno-biosensors, enzymatic biosensors, while with respect to the detection process we have electrical, optical, piezoelectric and electrochemical biosensors. The electro-chemical biosensors are widespread and further divided into impedimetric, poten-tiometric or amperometric sensors. All these categories of sensors are able to detect glucose, cholesterol, E-coli, viruses, cortisol, DNA amino acids and vapors and gases relevant for breath diagnosis, for example, or dangerous gases (Kuila et al. 2011). Graphene biosensors are able to detect the main cancer biomarkers, cancer cells and pathological bacteria and pathogens, traces of the Alzheimer disease, and can act as scaffolds for stem cells.

Graphene is considered as an elite nanomaterial for biosensing (Kumar et al. 2015). Among the many unique attributes of graphene, of relevance for this application is that in all carbon-based nanomaterials there is a direct electron transfer between the functionalized nanomaterial and the bioreceptor, without any mediator (fluorescent molecules, chemical groups, etc.). As such, graphene amplifies the output signal and confers high sensitivity to free-label sensing. Graphene is also a biocompatible material.

A large majority of graphene biosensors are based on graphene FETs and FET arrays, where the biomolecules to be sensed change the conductance of the channel, thus modifying the drain current and the transconductance. A typical graphene FET biosensor with liquid gate is depicted in Fig. 1.91.

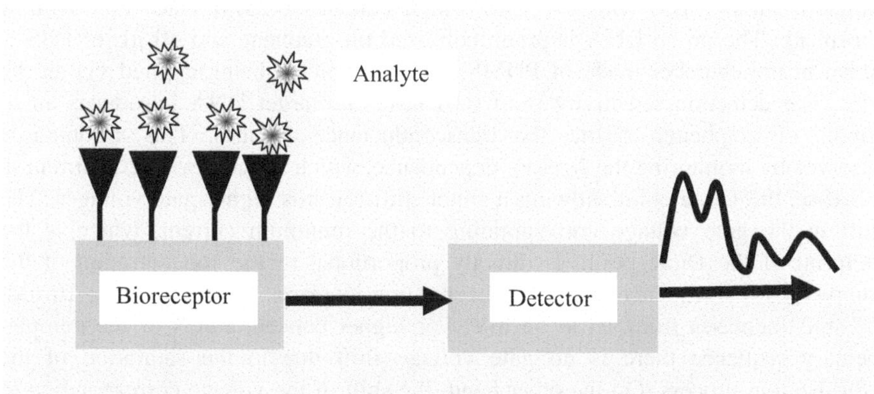

Fig. 1.90 Biosensor configuration

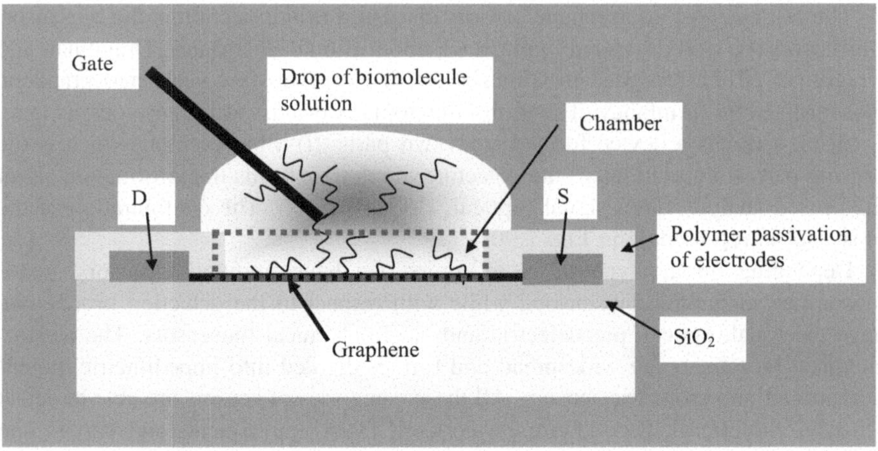

Fig. 1.91 Typical configuration of graphene biosensor based on graphene FET

Molecules are anchored to graphene, functionalized in a special manner to hold the majority of biomolecules on the graphene channel. Hence, significant changes in the channel conduction are produced by the sensed biomolecules. The main property of the graphene FET biosensor is the liquid gate. In principle, the ionic liquid separates the gate electrode and the graphene sheet that forms the transistor channel. At low gate bias, the current flow is limited by the formation of a Debye layer (a double ionic layer), but at high gate voltages the electrolytic processes produce relatively high currents between the gate and graphene. Graphene FETs based on liquid gates are also used to detect proteins, small molecules and viruses (Green and Norton 2015).

The ultimate performance of such a sensor was the detection of a single base mutation from a DNA sequence (Dong et al. 2010). The probe, the complementary DNA and the one-base mismatched DNA string were dissolved in phosphate buffered saline (PBS, with pH 7.4), which contains 0.25 M NaCl and 10 mM phosphate. The probe DNA is preimmobilized on graphene and 40 μL of PBS is added in the chamber made of PDMS, the gate voltage being applied via an Ag wire. The detection sensitivity is of 0.01 nM. The target DNA introduces an *n*-doping of graphene, so that the transconductance changes. This *n*-doping is observed by monitoring the $I_D - V_G$ dependence, which shifts upwards in current in this case, the Dirac point showing a slight shift towards higher gate voltages. The shift in the gate voltage corresponding to the minimum current, which is the footprint of the Dirac point, is directly proportional to the concentration of the complementary sequence. As this concentration increases from 0.01 up to 10 nM, the shift increases from 10 to 55 mV. For higher concentrations of the complementary sequence there is no gate voltage shift due to the saturation of the hybridization process. On the other hand, the shift of the voltage corresponding to minimum current is less pronounced for one-base mismatched DNA string, the shift being of only 20 mV for a concentration of 500 mM. The biosensing mechanism is displayed in Fig. 1.92.

Another important application of graphene in biosensors is DNA sequencing using nanopores, which implies the electrical detection of the sequence of bases A, G, C and T that form the DNA code. DNA sequencing provides important information for the understanding and early detection of serious diseases such as cancer. The electrical detection of DNA sequences is based on the fact that the DNA backbone is charged in ionic solution. More precisely, a single negative charge corresponds to each nucleotide, or two charges are associated to a pair of nucleotides for ds (double-stranded) DNA. As the negatively charged DNA is dragged by an applied electric field through a nanopore with a very small diameter, of 1–2 nm, the bases generate distinct electrical signals, which are further amplified and processed (Zwolak and di Ventra 2008). The nanopore diameter must be so small because the DNA diameter is no larger than 2 nm, the length of a nucleosite being of 0.7 nm. Thus, the limits of nanotechnologies and electronics are reached, and in many ways a nanopore-based DNA sequencing device, as that represented in Fig. 1.93, is one of the most advanced nanoelectronic circuits.

DNA sequencing using a nanopore consists of two ionic chambers with the roles of trapping and translocating the DNA sequence, respectively, followed by electronic detection. The ionic chambers that trap and translocate the DNA sequence are labelled as *cis* and *trans*. By applying a DC bias across the nanopore, a DC electric field is created, which generates an ionic current through the pore by pulling out the negatively charged DNA through the pore.

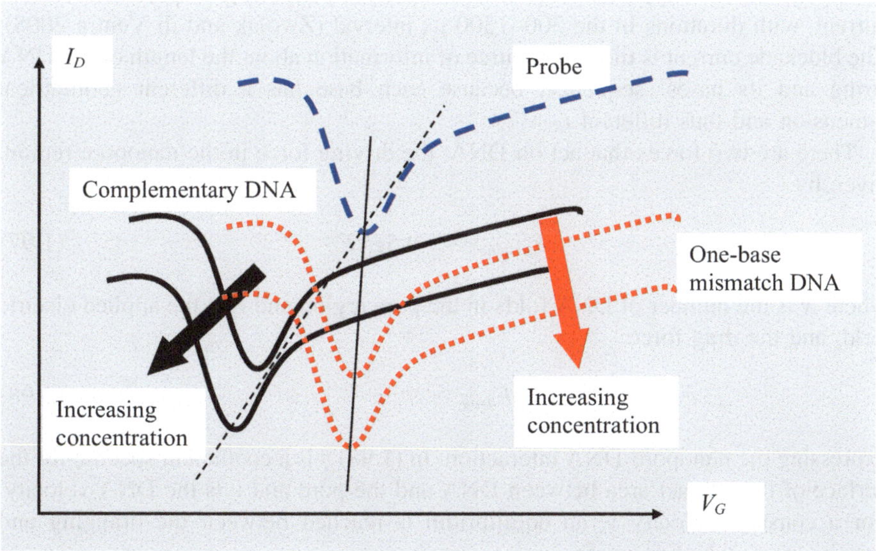

Fig. 1.92 DNA hybridization and one-base mismatch detection

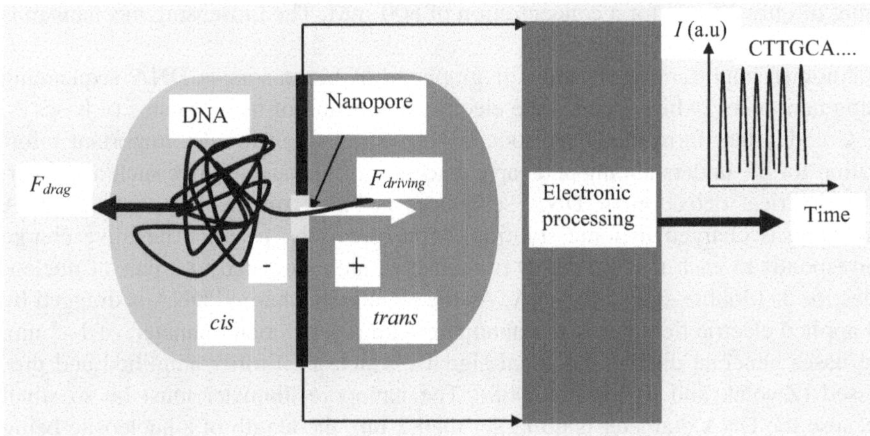

Fig. 1.93 DNA nanopore sequencing circuit

The trapping of DNA depends on the concentration of the solution and the strength of the applied DC electric field, while the translocation is dependent on the applied field (bias) and many other factors (viscosity, ionic concentrations, etc.), which are generally referred to as DNA-pore interaction effects. The physical principle of DNA sequencing with the help of a nanopore, shown in Fig. 1.94, is based on the fact that translocation occurs in a certain time duration t_d, and during this time the strand of DNA disables the flow of ions through the pore. Because the nucleotide localized in the pore has a much slower velocity than the ions, the DNA nucleotide inside the pore blocks the ionic current, inducing a spike in the ionic current, with durations in the 300–1300 μs interval (Zwolak and di Ventra 2008). The blockade current is the main source of information about the length of the DNA string and its bases' sequence, because each base has a different geometrical dimension and thus different t_d.

There are two forces that act on DNA: the driving force in the nanopore region, given by

$$F_{driving} \cong -0.5eNE \tag{1.97}$$

where N is the number of DNA folds in the pore region and E is the applied electric field, and the drag force

$$F_{drag} \cong N\gamma\, v \tag{1.98}$$

expressing the nanopore-DNA interaction. In (1.98) γ is a coefficient specific for the surface of the contact area between DNA and the pore and v is the DNA velocity. For a constant velocity v, an equilibrium is reached between the dragging and driving force, which requires

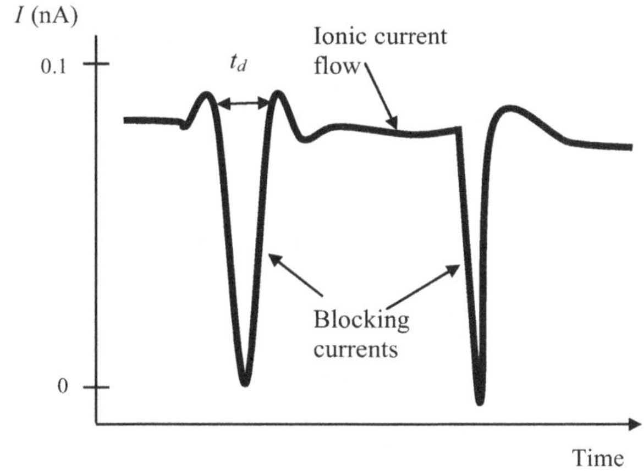

Fig. 1.94 A schematic behavior of the DNA blocking current

$$F_{drag} = F_{driving} \tag{1.99}$$

An enhanced sensitivity of DNA sequence detection is achieved in the transverse electronic current method, in which two DC biased electrodes are added on the two sides of the nanopore, as illustrated in Fig. 1.95, and the transverse current is monitored.

The parallel electric field, denoted by $E_{\|}$ in Fig. 1.95, drags the DNA through the nanopore, while the transverse field, $E_{\perp}$, contains the DNA sequencing information, since each base has a very distinct current signature (Zwolak and di Ventra 2008). More precisely, as suggested by Fig. 1.96, each nucleotide translocating through the nanopore has a distinct electronic signature of the bases that is imprinted in the

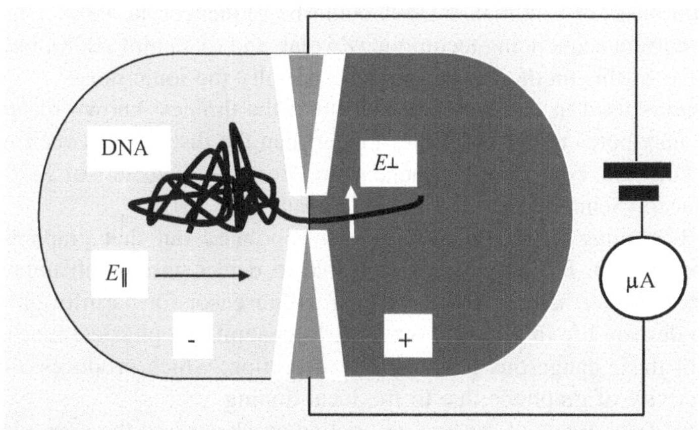

Fig. 1.95 DNA sequencing using the transverse current method

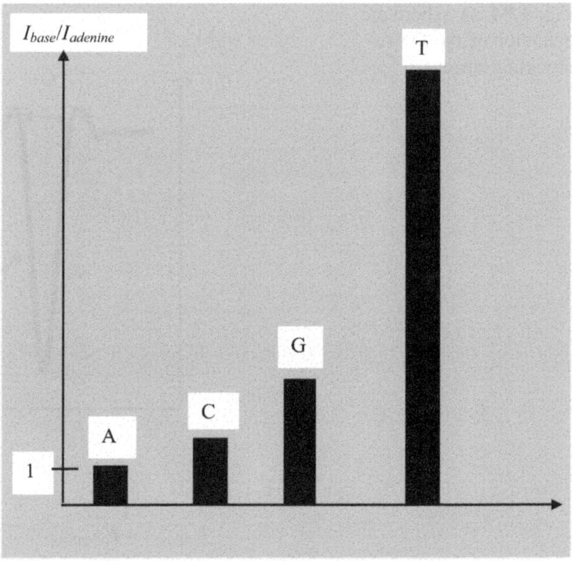

Fig. 1.96 Amplitudes of base currents

transverse current. The origin of this important distinction between the transverse currents corresponding to different DNA bases is the Fermi level position with respect to the HOMO and LUMO, and the DOS values, since the bases have different spatial extent and thus different extended molecular orbitals. The nanopore current (Zwolak and di Ventra 2008) is given by:

$$I \cong e^2 \Gamma_L(E_F)\Gamma_R(E_F)V/[\pi\hbar(E_F - E_N)^2] \quad (1.100)$$

where Γ_L and Γ_R are coupling coefficients, which are very different for each base due to their different shapes, corresponding wavefunctions and Fermi levels. The three billion bases of a human ssDNA could be sequenced in about 7 h using the transverse current sequencing technique (Zwolak and di Ventra 2008), but the main disadvantage of this method is the noise, especially the ionic noise.

Graphene is used in nanopores because it is the thinnest known material, other solid-state nanopores being 100 times longer than the distance between two bases, which is of 0.5 nm. Few-layer graphene nanopores with diameters of 2–20 nm have been produced (Schneider et al. 2010; Merchant et al. 2010).

At the beginning of this section we have pointed out that graphene and its various forms (GO, RGO, aerogels) are able to detect gases with unprecedented sensitivity. We refer here to the gas sensor as biosensor for harmful gases, which are able to destroy life in a very short time. In general, the physical mechanism for detection of these dangerous gases is gas adsorption, which produces a change in the conductivity of graphene due to the local doping.

There are many types of sensors on pristine graphene (see the examples above), graphene oxide (GO), reduced oxide graphene (RGO) or graphene aerogels

(Varghese et al. 2015). Pristine graphene gas sensors are so sensitive that they can detect the adsorption or desorption of a single gas molecule from the graphene surface (Schedin et al. 2007). The sensing mechanism is the change in graphene resistivity due to local doping induced by the gas molecules. In particular, graphene detection of NO_2 reaches an outstanding performance of 0.1 ppm.

Graphene sensors in a Hall bar configuration were used also to understand the type of doping induced by various molecules. It was found out that *n*-type doping is produced by NH_3, CO and ethanol, while *p*-type doping is induced by NO_2, H_2O and iodine. The detection limit of Hall bar gas sensor is 1 ppb, concentration at which the relative resistivity of graphene changes by $\Delta\rho/\rho \propto 10^{-4}$.

The decrease of contact resistance at 100 Ω in few graphene layers devices due to adsorption and desorption of a single NO_2 molecule is also used for sensitive detection. In this case, the difference in resistance, of 2.5 Ω, is very large considering that a single molecule is attached or detached from the graphene surface.

Other forms of graphene are able to detect harmful gases with a very high sensitivity and selectivity. For example, flakes of GO deposited on a Pt IDT electrode array by the drop casting method detect NO_2 with a sensitivity of 20 ppb (Prezioso et al. 2013). In this sensor, represented schematically in Fig. 1.97, detection is associated to the oxygen functional groups at the GO surface, on which NO_2 is mainly adsorbed.

A promising path for highly sensitive gas sensors is offered by 3D graphene aerogel-ZnO composites (Liu et al. 2015b). Semiconductor oxides, such as ZnO, are extensively used in gas sensors, but ZnO sensors work at high temperatures due to the high activation energy required for surface adsorption and redox reaction

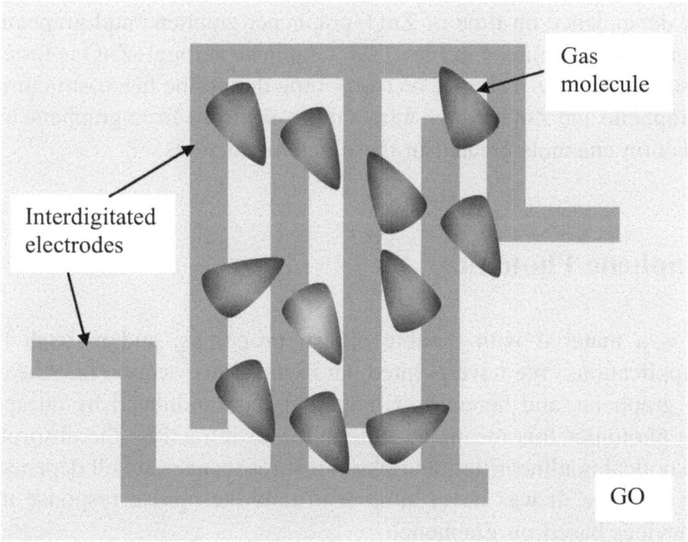

Fig. 1.97 GO gas sensor

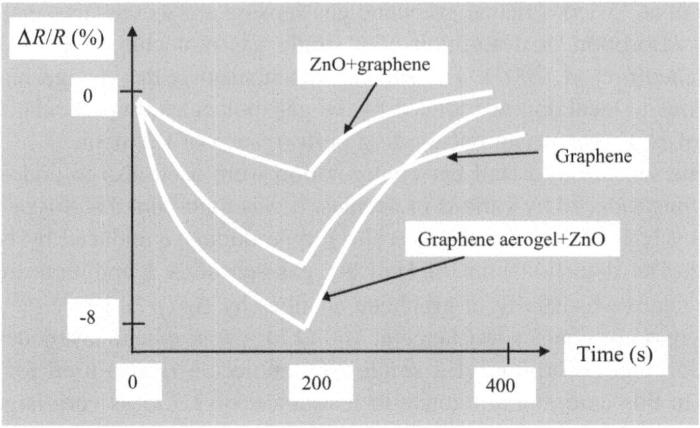

Fig. 1.98 Typical response in time of graphene aerogel+ZnO in comparison with other related sensors at 50 ppm NO_2 gases (after Liu X et al. 2015a, b)

between ZnO and the gas. On the other hand, ZnO spheres in combination with graphene aerogel display different physical sensing mechanisms, which do not require high temperatures. The graphene aerogel creates a conductive matrix between ZnO spheres, necessary for sensing and for avoiding also agglomeration between spheres. The sensing process as well is enhanced significantly due to the large specific surface of graphene aerogel. Thus, the graphene aerogel-ZnO composite works as gas sensor at room temperature, with fast response and recovery times. For example, the sensitivity of this sensor for NO_2 detection is 50 ppm at room temperature, the resistance variation $\Delta R/R$ at 200 ppm of NO_2 being of 20%. The $\Delta R/R$ dependence on time of ZnO+graphene, graphene and graphene aerogel and ZnO spheres is displayed in Fig. 1.98. Graphene aerogel+ZnO is faster than the graphene sensor and has a shorter recovery time due to the heterostructures formed between graphene and ZnO, which attract more electrons from graphene to NO_2 via new conduction channels created in these heterostructures.

1.4 Graphene Photonics

Graphene is a material with tunable optical properties, and as such has many photonic applications. We have pointed out in the above section that the density of carriers in graphene, and hence its conductivity, is modulated by an applied DC voltage. In photonics this means that the index of refraction, the absorption coefficient, the optical nonlinearities, and plasmonic resonances are all dependent on the applied DC voltage. It was never simpler to tune the optical response than using photonic devices based on graphene.

Intrinsic graphene has a universal optical conductance (Low and Avouris 2014)

$$G_{opt} = e^2/4\hbar \tag{1.101}$$

and thus an absorption

$$\alpha\pi = 2.3\% \tag{1.102}$$

where $\alpha = e^2/\hbar c \cong 1/137$ is the fine structure constant. Equation (1.102) shows that the absorption is independent of the wavelength of light, in strong contradiction with semiconductors, where absorption has a narrow band centered on the wavelength that corresponds to the bandgap.

The optical conductivity has a more general expression (Sensale-Rodríguez et al. 2013) and is dependent on the type of optical transition: intra- or interband, as illustrated in Fig. 1.99.

We have

$$\sigma_{intra} = ie^2 E_F/[\pi\hbar^2(\omega + i/\tau)] \tag{1.103a}$$

and

$$\sigma_{inter} = (ie^2\omega/\pi)\int_0^\infty dE[f(E - E_F) - f(-E - E_F)]/[(2E)^2 - (\hbar\omega + i\Gamma)^2] \tag{1.103b}$$

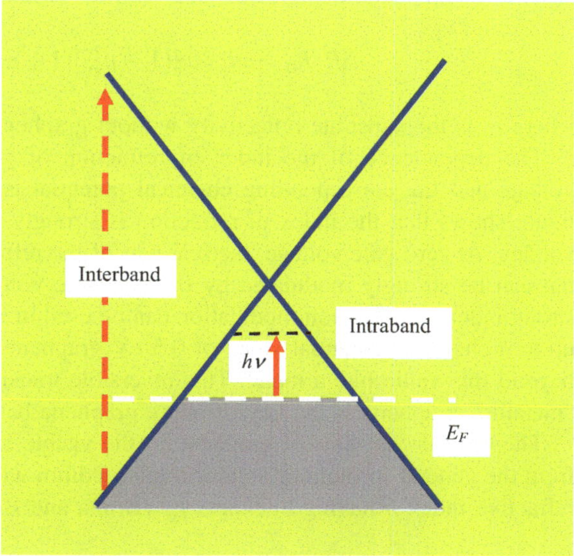

Fig. 1.99 Graphene optical transitions

where τ and Γ are the momentum relaxation time and broadening of interband transitions, respectively, and $f(E) = 1/[(\exp(E - \mu_c)/k_B T) + 1]$ is the Fermi distribution, with μ_c the chemical potential describing charge accumulation in graphene. This latter parameter is related directly to the applied voltage V via

$$\mu_c = \hbar v_F \sqrt{(\pi \varepsilon_{ox}/e d_{ox})(V - V_{Dirac})} \tag{1.104}$$

where ε_{ox} is the permittivity of dielectric, d_{ox} its thickness and V_{Dirac} the Dirac voltage.

Because the Fermi level is tunable via the carrier density: $E_F = (\pi n)^{1/2}/h$, which in turn is changed via the electrostatic doping induced by the gate voltage, the optical conductivity is DC tunable. So, via the Fermi energy shift and the chemical potential dependence on the applied voltage, the optical absorption in graphene depends on the DC voltage since σ is directly proportional to absorption.

When the optical frequency is very high, including the visible, UV, and far IR ranges, the conductivity becomes $e^2/4\hbar$, independent of frequency.

Another consequence of the direct proportionality between σ and absorption is that the index of refraction becomes dependent on the gate voltage via the formula:

$$n_{gr}(V_G) = \sqrt{1 - i\sigma(V_G)/(\omega \varepsilon_0 t_{gr})} \tag{1.105}$$

where $t_{gr} = 0.34$ nm is the thickness of graphene. The dependence $n_{gr}(V_G)$ was recently experimentally proven (Xu et al. 2015) at 1550 nm. In these experiments graphene was transferred on a Si/SiO_2 substrate, where the doped Si substrate played the role of the backgate. The measurements were performed in a reflection configuration, the change in the power reflectivity with respect to the value R_0 in the absence of graphene being calculated with the formula

$$\Delta R/R_0 = \pi t_{gr} \mathrm{Re}[(1 - n_{gr}^2)(1 + r_0)^2/(r_0 \lambda)] \tag{1.106}$$

where r_0 is the substrate reflectivity without graphene.

The dependence of the index of refraction of graphene on the applied gate voltage and the corresponding chemical potential is depicted in Fig. 1.100. This figure shows that the index of refraction is strongly tunable with the applied gate voltage. At zero gate voltage, the real part of the refractive index of graphene is 3, but can be strongly modulated by the backgate voltage even below 1. This very strong index of refraction modulation can be used in several photonic devices since up to a chemical potential value of 0.5 eV graphene behaves like a dielectric and beyond this value like a metal. This reversible transition between "dielectric" and "metallic" graphene is the key of many graphene-based modulators.

The optical constants of graphene in the visible spectral range can be deduced from the general formula of an absorbing medium with extinction coefficient k and refractive index n having thickness t_{gr} (Bruna and Borini 2009):

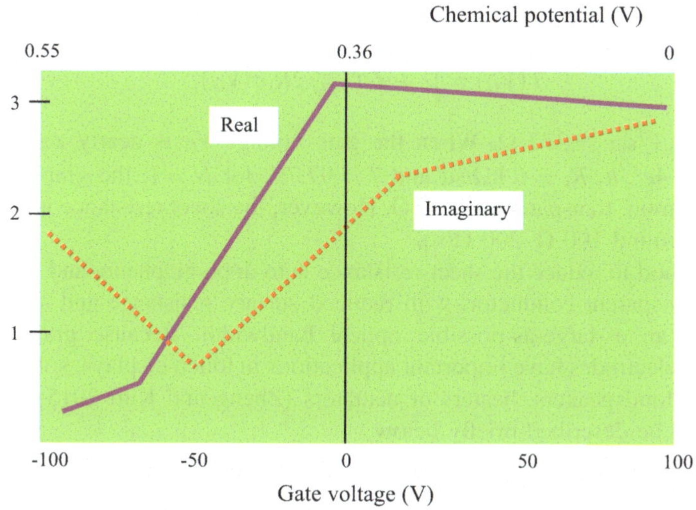

Fig. 1.100 Graphene refractive index as a function of the gate voltage (after Xu et al. 2015)

$$4\pi k/\lambda = -(nt_{gr})^{-1}\ln[(I/I_0)(1-R)^{-1}] \tag{1.107}$$

Because in the graphene monolayer

$$I/I_0 \approx (1-\alpha\pi) \tag{1.108}$$

and $1 - R \cong 1$, we obtain

$$k = -(\lambda/4\pi n t_{gr})\ln(1-\pi\alpha) \equiv C_1\lambda/n \tag{1.109}$$

where $C_1 = 5.446\ \mu\text{m}^{-1}$. This is a simple relation between the index of refraction and the extinction coefficient. Assuming that n is constant in the visible domain, $n = 3$, it follows that $k = C_1\lambda/3$.

The optical transmission and reflection in graphene are also tunable by the gate voltage. The optical transmission depends on the optical conductance G_{opt} as (Bonaccorso et al. 2010):

$$T = [1 + G_{opt}N/(2\varepsilon_0 c)]^{-2} \tag{1.110}$$

where N is the number of graphene layers. Because the sheet resistance R_s is determined by the DC gate-dependent conductivity via:

$$R_s = [\sigma(V_G)N]^{-1} \tag{1.111}$$

it follows that

$$T(V_G) = [1 + Z_0 G_{opt}/2R_s\sigma(V_G)]^{-2} \tag{1.112}$$

where $Z_0 = 1/\varepsilon_0 c = 377\ \Omega$. When the gate voltage V_G is nearly zero, the conductivity is $4e^2/h$, $R_s = 6$ kΩ/sq and $T = 97.7\%$ for $N = 1$, the graphene monolayer being more transparent than ITO. However, the sheet resistance is too high; it should be around 100 Ω–200 Ω/sq.

One method to reduce the sheet resistance is to dope graphene, and to transform it into a transparent conductor, with reduced surface resistance and a high transmission in an as-large-as-possible optical bandwidth. Because graphene-based transparent electrodes have important applications in touch displays, solar cells, and transparent loudspeakers, heaters or actuators (Zheng and Kim 2015) the doping process will be described briefly below.

We have already discussed about graphene doping in several contexts. First, the role of doping is to engineer the Fermi level in graphene, and thus to control the charge carrier density. By shifting the Fermi level away from the Dirac point, we are able to decrease the sheet resistance of graphene. The doping could be p-type or n-type, depending on the doping method. We have already mentioned the electrostatic doping of graphene, which is one of the most important properties of graphene and 2D materials, and in which the Fermi level is shifted by a gate voltage, but electrostatic doping is not applicable for transparent electrodes. Other doping methods were mentioned in connection with graphene functionalization. Here we discuss the main methods of doping for transparent electrodes.

The wet doping is referred to as spin coating or dip coating of graphene in a solution. Acid treatment is a typical doping method, and the nitric acid (HNO_3) is used extensively for this purpose, produces a p-type doping and a shift of the Fermi level of up to 0.8 eV depending on its concentration (see Fig. 1.101). Halogenating agents such as $SOCl_2$, produce also p-type doping of graphene and reduces the sheet resistance with 50–80%. Auric chloride $AuCl_3$ is also a p-type dopant. In the case of multilayer graphene, the sheet resistance decreases with 80% as a result of this doping method, becoming less than 100 Ω/sq. Note that the optical transmittance is preserved beyond 90%.

Another method of doping is hybridization with carbon nanotubes, metal nanowires, nanorods, and metal grids. For a good review of doping methods applied for graphene as transparent conductor see Zheng and Kim (2015, Chap. 4).

Nonlinear effects in graphene are very strong and are caused by the interaction between optical fields and electrons on the outer shell of carbon atoms, which shifts the electron cloud from its initial position around the nucleus and generates a net polarization. At moderate optical fields, there is a linear relation between the polarization P and the electric field E (Bao and Loh 2012)

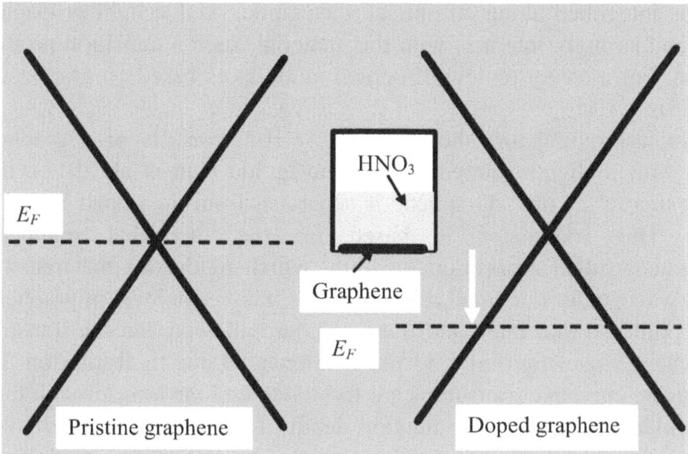

Fig. 1.101 Graphene doping using nitric acid

$$P = \varepsilon_0 \chi^{(1)} E \tag{1.113}$$

where $\chi^{(1)}$ is the first order linear susceptibility. At stronger optical fields, however,

$$P = \varepsilon_0 (\chi^{(1)} + \chi^{(2)} E^2 + \chi^{(3)} E^3 + \ldots) \tag{1.114}$$

where $\chi^{(2)}$ and $\chi^{(3)}$ are the second- and third-order nonlinearities, respectively, which are related to nonlinear phenomena such as saturable absorption, self-focusing or two-photon absorption. Graphene displays very large third-order nonlinearity, of 10^{-7} esu, and a very large Kerr coefficient and optical damage threshold, with orders of magnitude larger than in semiconductors (see Table 1.6).

In relation (1.113), the real part of $\chi^{(1)}$ is associated to the real part of the refractive index, while the imaginary part is linked to optical loss or gain. Graphene modulators are based on the tunability of the real part of $\chi^{(1)}$ with an applied DC field, i.e. the variation of the index of refraction with the DC voltage. We have seen above that when the gate voltage, i.e. the chemical potential, varies graphene undergoes a reversible transition from a "dielectric" to a "metallic" state. If

Table 1.6 Nonlinear optical properties of graphene and other materials (Bao and Loh 2012)

Material	Refractive index	Optical damage threshold (MW/cm^2)	Third order nonlinearity (esu)	Kerr nonlinear coefficient at 1.55 μm
Graphene	2.6–3	3×10^6	10^{-7}	10^{-11}
h-BN	2.2	500	1.36×10^{-14}	–
GaAs	3.4	45	4×10^{-8}	3.3×10^{-17}
Si	3.44	500	5×10^{-11}	4.5×10^{-18}

graphene is integrated along an optical waveguide, so that light propagates along graphene and strongly interacts with this material, such a transition produces light modulation. For a recent review of optical modulators based on graphene see Luo et al. (2015).

The simplest modulator, shown in Fig. 1.102, consists of a graphene sheet integrated with a slightly doped Si bus waveguide (Liu et al. 2011), the device having an area of 25 μm^2. Graphene is separated from the doped Si by an Al_2O_3 dielectric. This modulator is based on the fact that graphene is a DC-voltage-controlled absorbtion medium, which modulates the transmission of the Si bus waveguide. For small drive voltages, the Fermi level in graphene is near the Dirac point, so that interband transitions are allowed and the transmission of photons with a wavelength of 1.53 μm is attenuated due to absorption. For larger drive voltages, interband transitions are forbidden and the transmission through the bus waveguide increases. A modulation depth of 0.1 dB/μm was achieved in this way in the bandwidth 1.35–1.6 μm, the operation bandwidth being of about 1 GHz.

A similar double-layer graphene modulator operating around 1 GHz has an enhanced modulation depth, of 0.16 dB/μm, at a drive voltage of 5 V. The entire graphene-Al_2O_3 dielectric-graphene structure is functioning as a *p-i-n* junction to minimize the losses (Liu et al. 2012). This type of modulator is schematically represented in Fig. 1.103. The state-of the-art of optical modulators based on Si and graphene using the concepts described above is a broadband 10 Gb/s modulator at telecom wavelength (Hu et al. 2014).

A double layer formed from two graphene monolayers separated by a dielectric in a FET-like configuration and integrated on a Si waveguide is able to work simultaneosly as modulator and photodetector (Youngblood et al. 2014). In this device, illustrated in Fig. 1.104, the bottom graphene monolayer acts as a channel, playing also the role of an optical absorber where photocarriers are generated, while the top graphene electrode is an optical transparent gate able to tune the optical and electrical behavior of the graphene channel. The thickness of the Al_2O_3 gate dielectric is 100 nm. The drain and source contacts are made from different metals,

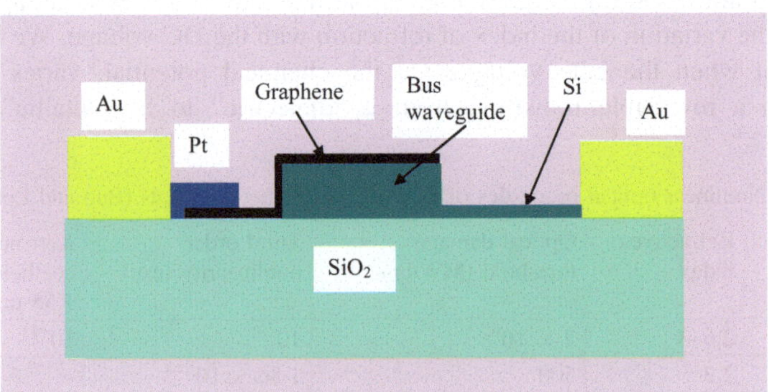

Fig. 1.102 Optical modulator based on a slightly doped Si waveguide

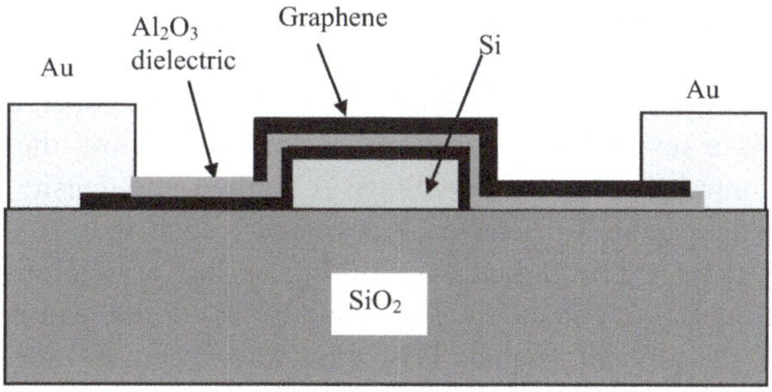

Fig. 1.103 Optical modulator based on double-layer graphene

Ti/Au and Pd/Au, with different workfunctions, these electrodes doping the graphene channel *n*-type and, correspondingly, *p*-type. As a result, if the middle of the graphene channel is tuned to the Dirac point, the entire device mimics a lateral *p-i-n* junction, so that a built-in field located in the channel separates the photocarriers. The device displays a modulation depth of 64% in the GHz bandwidth and a responsivity of 57 mA/W.

Another photonic device is presented in Fig. 1.105, where a graphene–graphene capacitor is integrated with a silicon nitride ring resonator coupled to a bus waveguide. The capacitor, consisting of two graphene monolayers separated by 65 nm Al_2O_3 dielectric, is grown on top of a portion of the ring resonator. When both graphene sheets are unbiased, the bus waveguide and ring resonator, with a 300 nm gap between them, are practically decoupled due to the high losses in the ring caused by opaque graphene; light propagation through the bus waveguide is almost unimpeded. On the other hand, when a DC bias is applied on the graphene capacitor, the transmission changes dramatically, from −8 to −25 dB at –50 V, so that light propagation through the bus waveguide is obstructed. In this case, graphene is electrostatically doped, the shift of the Fermi level preventing light absorption, and the ring resonator and bus waveguide become coupled. Light modulation is caused by graphene-tuned interference in the coupling region between the ring and the bus waveguide. The device has 30 GHz bandwidth operation, a 15 dB tuning of cavity transmission being achieved with only 10 V swing (Phare et al. 2015).

All the above optical modulators are based on interband transitions, but modulation at higher frequencies, up to 3 THz, is possible using intraband transitions. The

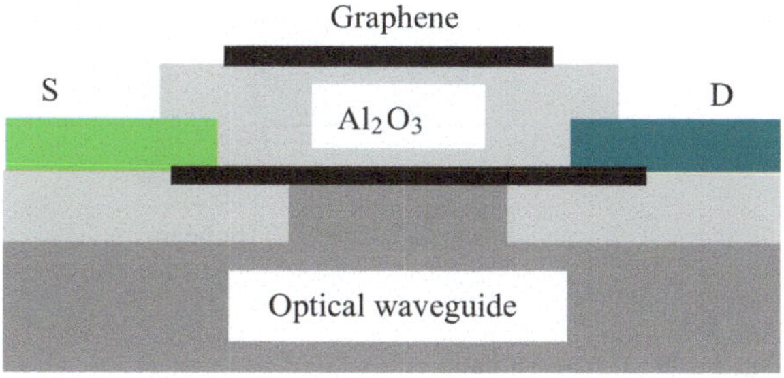

Fig. 1.104 Modulator and photodetector based on graphene

working principle of such modulators is similar to that of optical modulators in the visible or IR spectral regions. Graphene at the Dirac point is transparent, having zero charge density and behaving like a "dielectric" such as ITO in the visible spectrum. As a result, the absorption has low values and the transmission $T \to 1$. On the contrary, when a gate voltage is applied, the density of carriers increases and graphene tends to behave as a "metal", such that $T \to 0$ (Sensale-Rodríguez et al. 2012). A THz modulator working on this principle is represented schematically in Fig. 1.106. The modulation depth is 64% with the insertion loss of 2 dB at 0.62 THz.

All-optical spatial THz modulators can be successfully implemented using graphene transferred on Ge. Ge was chosen as substrate instead of Si due to its higher bulk mobility, and thus an increased surface mobility and better modulation performances. In this graphene-based modulator, a wideband modulation, from 0.25 up to 1 THz, was obtained under laser pumping at 1550 nm (Wen et al. 2014).

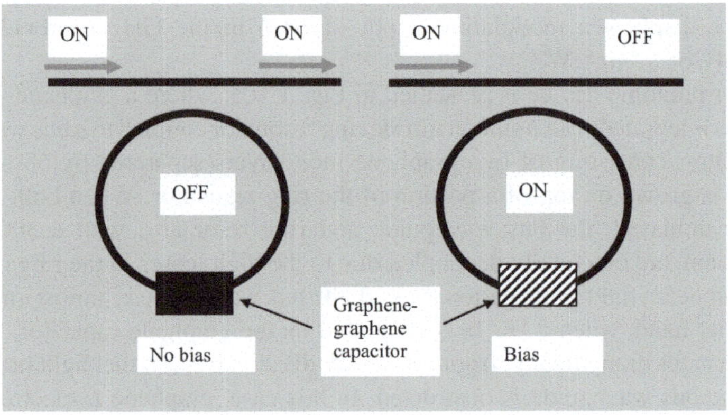

Fig. 1.105 Graphene modulator with a bandwidth up to 30 GHz

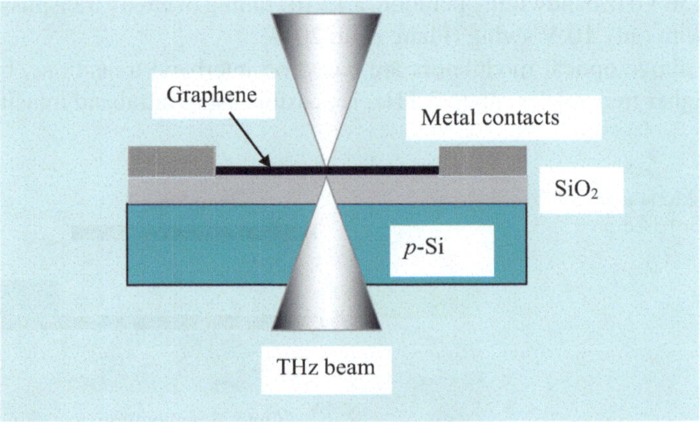

Fig. 1.106 THz modulator

The modulation depth is impressive, reaching 94%, and the measured modulation speed is 200 kHz at a 340 GHz carrier. The modulator is represented schematically in Fig. 1.107 and its modulation principle is described in Fig. 1.108. No bias needs to be applied on graphene for modulation purposes in this case. Instead, at a laser pumping power of 400 mW, both graphene and Ge layers show a very large attenuation of the THz wave due to the photogenerated free carriers, which form a thin conductive layer. But, in dark, i.e. when the optical power is reduced to 0, the transmission of THz waves reaches 50%. In this way, THz waves are modulating the optical carrier at 1550 nm.

The transmission is given by

$$T(E) = [1 + 0.5\alpha\pi\sigma(E)]^{-2} \tag{1.115}$$

where α is the fine structure constant and the nonlinear conductivity can be expressed as

$$\sigma(E) = \sqrt{(1 + \sigma_3(E)H_2)^2 + \sigma_3^2(E)H_1^2} \tag{1.116}$$

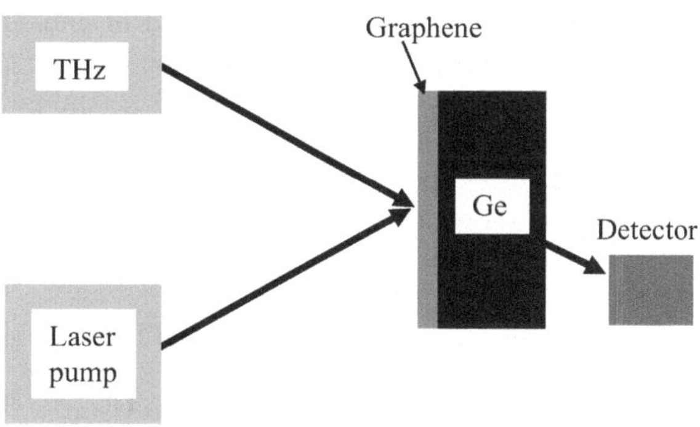

Fig. 1.107 All optical THz modulator

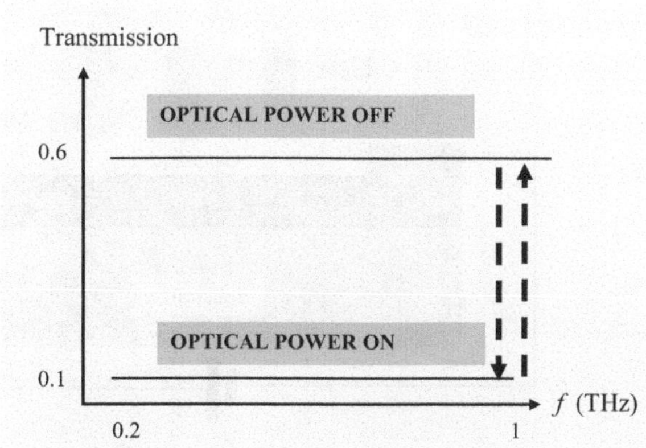

Fig. 1.108 Principle of the all-optical THz modulator

where

$$\sigma_3(E) = e^2 v_F^2 E_0^2/\hbar^2\omega^4,\ H_1 = (13/48)N(1/2) - (2/3)N(1) + (45/48)N(3/2),\ H_2 = 2N(1), \quad (1.117)$$

E_0 is the strength of the incident field, $N(x) = \tanh(x\hbar\omega/2k_BT)$ and T is the temperature; the terms H_1 and H_2 originate from the time averages of the nonlinear currents of the first and third order, respectively, in graphene. This model is precise enough to fit the experimental data.

Photodetection in graphene is due to interband and intraband absorption and generation of photocarriers. Since graphene absorption is independent of wavelength, this material can be used for photodetectors in a wide range of wavelengths, from microwaves up to visible, infrared and even ultraviolet. Various types of graphene, such as inks, flakes, or aerogels can be used for photodetection in specific types of devices, but there are also photodetectors fabricated at the wafer scale. The most common device for photodetection is the graphene FET.

There are several physical mechanisms for light detection using graphene (Sun and Chang 2014), the photoelectric effect being mostly used for this purpose. According to the photoelectric effect, excitons are formed when photons excite the graphene monolayer, which are separated in photoexcited carriers (electrons and holes) by the built-in field at the graphene-metallic electrode interfaces. The basic metal-semiconductor-metal-like (MSM-like) configuration of a graphene photodetector is shown in Fig. 1.109. Different configurations for electrodes, such as interdigitated electrodes, can be used, the substrate can be doped and then can act as the backgate, a top gate with the corresponding biasing circuit can be added to the configuration in Fig. 1.109, but in principle this is the basic configuration for graphene photodetection.

The photocurrent in graphene is given by

$$I_{ph} = e\mu A V \Delta n \quad (1.118)$$

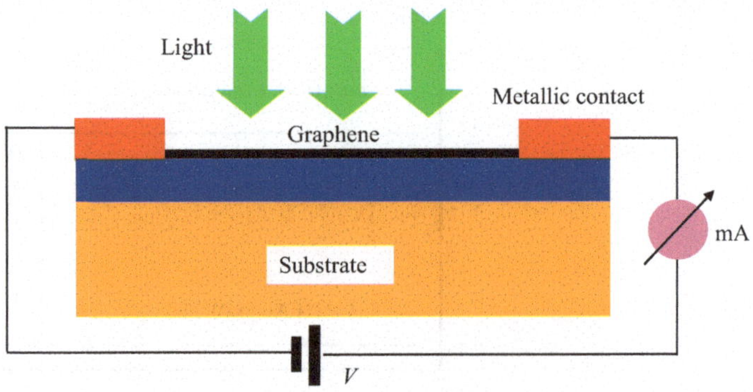

Fig. 1.109 MSM-like basic configuration of a graphene photodetector

where A is the cross section, V is the bias voltage, μ is the mobility of carriers, and Δn is the density of photocarriers. The carrier mobility in graphene is high, but the low absorption and the short lifetime of photocarriers generate low quantum efficiency in graphene and, despite many efforts to alleviate these issues, there are other 2D materials such as MoS_2 where the quantum efficiency is much higher. Therefore, graphene photodetectors have average performances and their most important advantage is the large bandwidth. The responsivities of graphene photodectors are below 10 mA/W, which is a modest performance taking into account that photodetectors based on 2D transition metal dichalcogenides have responsivities of 10^6 A/W and even higher.

Graphene photodection in a very large spectral bandwidth: ultraviolet, visible and near infrared, was demonstrated with graphene inks (GI) or GI functionalized with Au, Ag, or Au nanoparticles further encapsulated with bovine serum albumin (BSA) and deposited on interdigitated electrodes (IDTs) fabricated on a silicon dioxide/silicon substrate (Radoi et al. 2010). In contrast to gold-functionalized GIs, which have responsivities better than 1 mA/W at a 0.1 V bias over the huge bandwidth extending from 215 to 2500 nm, Ag-functionalized inks show at least a four-fold increased responsivity, with a record value of 13.7 mA/W in near infrared.

The IDT structures are fabricated on 0.5-μm-thick SiO_2 grown over a high resistivity silicon wafer ($\sim$8 kΩ cm). The Ti/Au IDT electrodes are fabricated from Ti with a thickness of 60 nm, the gold thickness being 240 nm. The width of an IDT electrode is 3 μm and the gap between adjacent digits is 1 μm. The SEM photo of 24 IDTs electrodes is displayed in Fig. 1.110.

Figure 1.111 illustrates the response of GI at white light. For the UV-VIS spectrum, the photodetector was illuminated with a halogen lamp terminated with

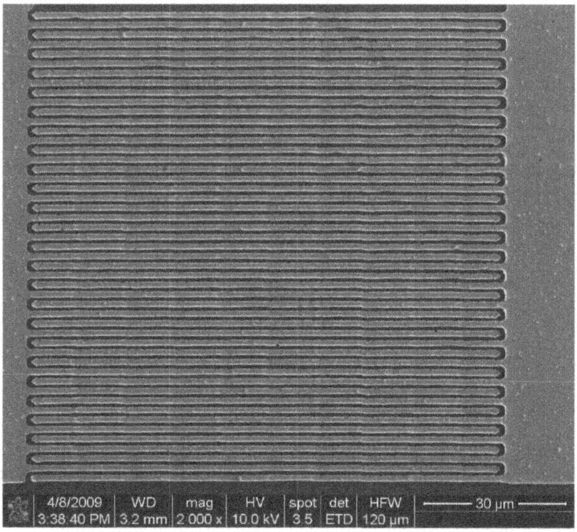

Fig. 1.110 SEM photo of graphene ink photodetector (from Radoi et al. 2010)

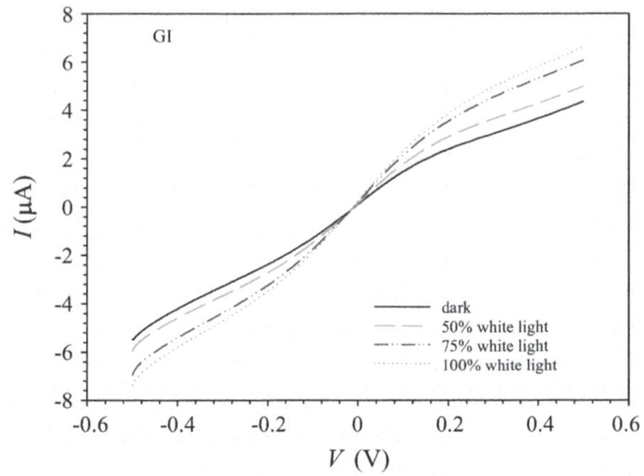

Fig. 1.111 GI response at white light with various intensities (from Radoi et al. 2010)

an optical fiber, having a power of 43 μW for the visible (VIS) spectral domain and a deuterium lamp with a power of 7 μW for ultraviolet (UV), both lamps covering the spectrum 215–1500 nm. For near-infrared (NIR), a tungsten halogen lamp with a power of 0.5 mW was employed, for an extension of the optical spectrum from 1500 to 2500 nm. Thus, all optical lamps cover the huge spectral spectrum from 215 up to 2500 nm. The optical responsivities of GI and GI functionalized with nanoparticles from UV up to FIR are displayed in Fig. 1.112.

There are many reviews explaining how the low absorption of graphene can be increased using optical cavity integration, optical waveguide integration, or plasmonic excitations (Koppens et al. 2014). These methods are well-known solutions for improving the performances of photodetectors of many types, so that we will not present them in detail here. But, as we have pointed out, graphene FETs act as electromagnetic detectors in the THz region when integrated with an antenna, and in the IR region if the transistors are integrated with antennas or excited via waveguides (Li et al. 2014).

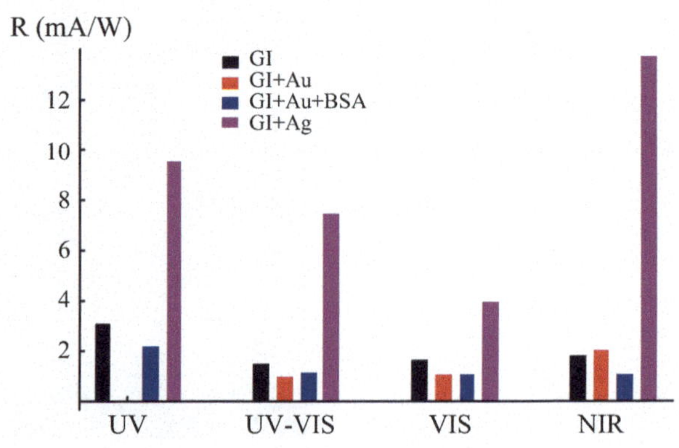

Fig. 1.112 Responsivities of graphene inks over different spectral regions at the constant bias of 0.1 V (from Radoi et al. 2010)

The best performance graphene photodetector is obtained by band structuring, which introduces electronic trapping centers and induces a bandgap in graphene (Zhang et al. 2013). In this way, a responsivity of 8.61 A/W can be achieved, which is about three orders of magnitude higher than that of the majority of graphene photodectors. This graphene photodetector is based on the physics of graphene; it does not just mimic (without much success) existing detecting solutions for other semiconductor photodetectors.

More precisely, a very thin Ti layer (2 nm) deposited on graphene and then removed after fabrication of a graphene dot-like array (see Fig. 1.113) induces a defect midgap states band (MGB) and a bandgap. The thickness of the Ti layer controls the average size of the graphene quantum dots. Thicker Ti layers produce smaller sizes of quantum dot-like features. For example, for a Ti thickness of $t =$ 20 nm, the average dimension of the quantum dots is $W = 7.5$ nm, while for $t = 2$ nm the average dimension increases to 20.5 nm. However, smaller features generate higher bandgaps, according to the formula

$$E_g \cong \hbar v_F / W, \tag{1.119}$$

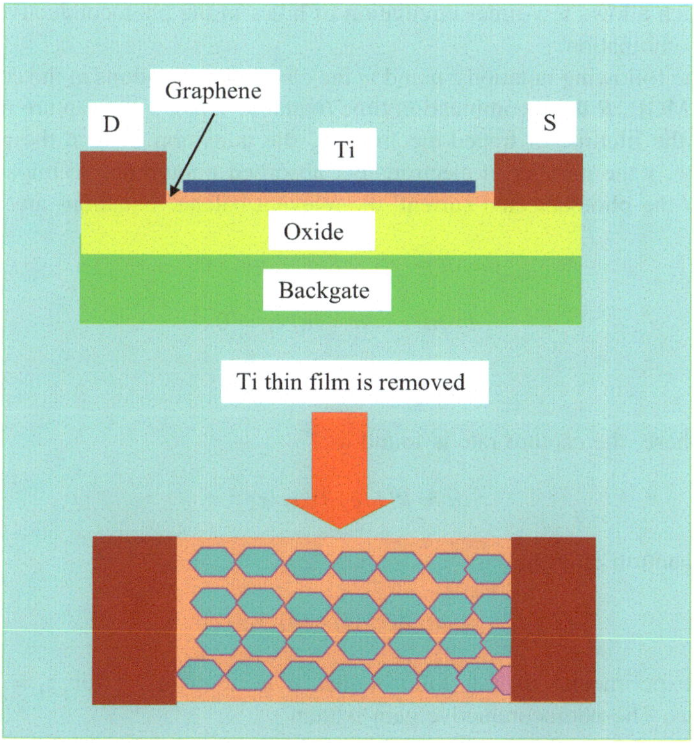

Fig. 1.113 Photodetector based on a graphene quantum dot-like array

while the average crystallite size of the quantum dots, L_c, are determined from the Raman analysis via the ratio of the intensities of the D and G bands, denoted by I_D and I_G, respectively:

$$L_c \cong (2.4 \times 10^{-10})\lambda^4(\mathrm{nm})(I_D/I_G)^{-1}. \tag{1.120}$$

There are two mechanisms competing for a high photoresponse

(i) impact ionization in the quantum-dot arrays; a photoexcited electron with a high energy relaxes to lower levels in the conduction band, exciting (one or few) carriers from the valence band into the conduction band;
(ii) recombination of an electron from the conduction band with a hole in the valence band, associated with the excitation of an electron in the conduction band to a higher energy state based on the resulting energy in the recombination process (Auger recombination).

Impact ionization is more efficient than Auger recombination in confined quantum systems, such as quantum dots, and thus carrier multiplication occurs, i.e. more and more electrons are produced in the conduction band. This mechanism is called multi-excitation generation (MEG). The photoconductive gain is caused in this case by trapping of the photogenerated and the secondary electrons on MGB states, which allows few-times circulation of holes in the photoconductive channel before recombination.

With the following notations: n and m the electron populations in the conduction band and MGB, R the recombination time (equal to 1 ps), α the capture rate of the MGB, τ_1 the lifetime of traped electrons, τ_t the transition time, β the photogeneration rate, χ the number of electrons per absorbed photon due to impact ionization, and I the photodetected current, the relevant balance equations are:

$$dn/dt = \chi\beta - Rm - \alpha n = 0 \tag{1.121a}$$

$$dm/dt = \alpha n - m/\tau_1 = 0 \tag{1.121b}$$

$$n + m = I\tau_t/e \tag{1.121c}$$

From these, the capture rate is found as

$$\alpha \approx R(I/e\chi\beta)(\tau_t/\tau_1) \tag{1.122a}$$

and the quantum efficiency is

$$\eta = \alpha n/\beta \cong (I/e\chi\beta)(\tau_t/\tau_1). \tag{1.122b}$$

From experimental data it follows that η is about 10^{-6} for τ_1 = 30 s and τ_t = 250 ns. The photoconductive gain is then

$$G = \eta(\tau_1/\tau_t). \tag{1.123}$$

This parameter depends on the energy of photons, and reaches 100 for a photon energy of about 2.3 eV.

The ultimate graphene photodetector, the configuration of which is represented in Fig. 1.114, is obtained by integrating the photodetector with an optical antenna, which works at visible and IR wavelengths. As an example, an antenna coupled to a graphene photodetector shows a responsivity of 17 nA/μW at 580 nm and has a signal-to-noise ratio (S/N) of 200 (Chakraborty et al. 2014).

In order to avoid the main drawbacks of graphene photodetectors: low absorption and short lifetime of photocarriers (of about 1 ps), mid infrared graphene detectors can be integrated with antennas. In such a configuration, illustrated in Fig. 1.115, the role of the metallic antenna array is to improve light absorption and to enhance the yield of photocarrier collection by increasing the photocarrier lifetime (Yao et al. 2014). The antenna array is formed from end-to-end coupled antennas and has a resonance in mid-infared, at 4.45 μm.

As a result of integrating the antenna array, the responsivity of the graphene photodetector increases to 0.4 V/W, which is orders of magnitudes higher than the same parameter in graphene photodetectors without antenna (2 mV/W). The gap size g between antennas is about 100 nm. The antennas are made of Pd/Au, the Pd layer, with a thickness of 10 nm being used to decrease the contact resistance.

The photoconductor current in the device in Fig. 1.115 is given by

$$I_{ph} = (eP_{inc}/h\nu)(2M\tau_R/\tau_{tr})[1 - \exp(-\tau_{tr}/\tau_R)] \tag{1.124}$$

where τ_R is the recombination lifetime, $\tau_{tr} = g/v_{tr}$ is the transit time across the gap and M is the multiplication factor of hot carriers. An equivalent circuit of the

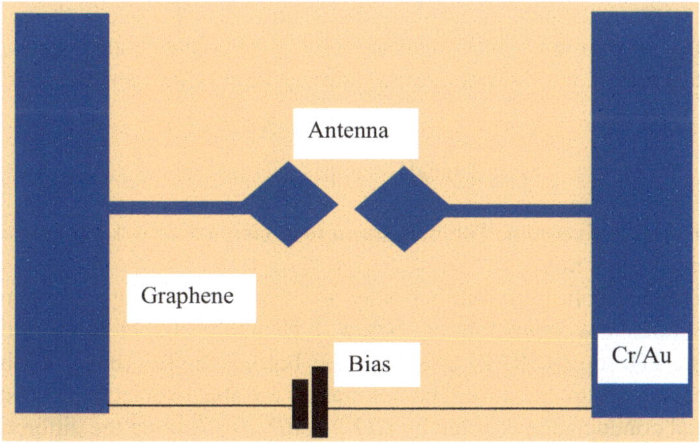

Fig. 1.114 Graphene photodetector antenna

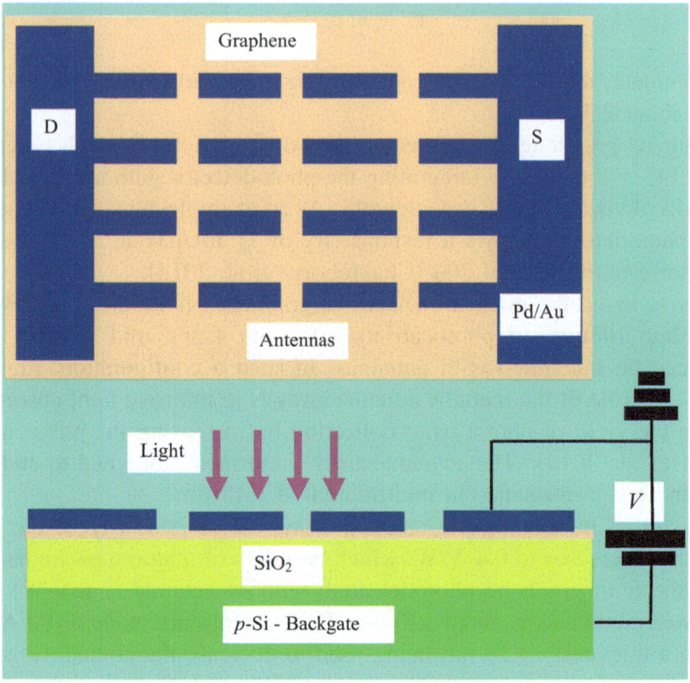

Fig. 1.115 Mid-infrared graphene photodetector: top view (*top*) and side view (*bottom*)

photodetector is presented in Fig. 1.116. Using this circuit, the responsivity of the photodetector can be defined as

$$R = (\alpha eG/hv)R_G/(R_G + R_S + R_L) \tag{1.125}$$

where R_G is the resistance of graphene between electrodes, R_S is the series resistance, accounting for the contact resistance between graphene and metals, R_L is the load resistance, α is the fraction of the incident light that is absorbed in graphene and

$$G = 2M\tau_R[1 - \exp(-\tau_{tr}/\tau_R)]/\tau_{tr} \tag{1.126}$$

is the photoconductive gain. The maximum responsivity, of 0.4 V/W, is attained at a drain current of 4 mA.

Bolometric detection is another way to sense electromagnetic waves. This method requires low temperatures, below 1 K, and is targeting electromagnetic waves from THz up to IR. In principle, the bolometer absorbes one photon and detects the temperature increase. The bolometers are able to detect the power P, since their thermal conductance is given by $G(T) = dP/dT$. Among the different ways of readout of a graphene bolometer, we mention the measurements of Josephson noise,

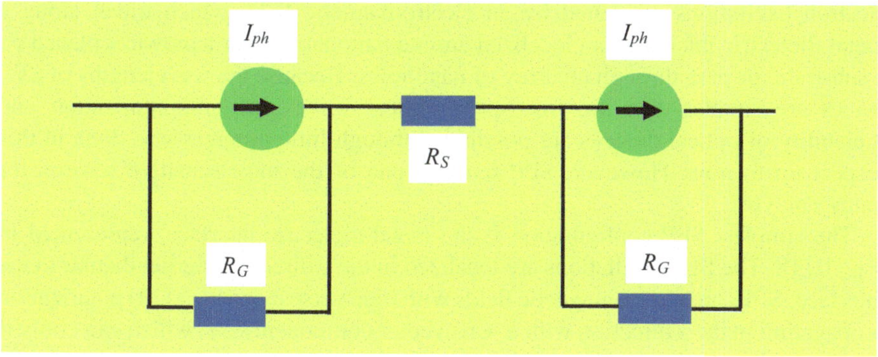

Fig. 1.116 Equivalent circuit of the graphene antenna photodetector

superconducting critical currents, or resistance of a superconducting tunneling contact. The working principle of the graphene bolometer is depicted in Fig. 1.117.

The figure of merit of a bolometric graphene detector is the energy resolving power, given by

$$R = E/\delta E \tag{1.127}$$

where E denotes the photon energy and δE the energy width. More details about graphene bolometers are found in Du et al. (2014).

The last issue related to graphene photonics concerns surface plasmon polaritons (SPPs) in graphene, which is a subject extensively treated in literature due to biosensing applications (Dragoman and Dragoman 2012). SPPs are collective

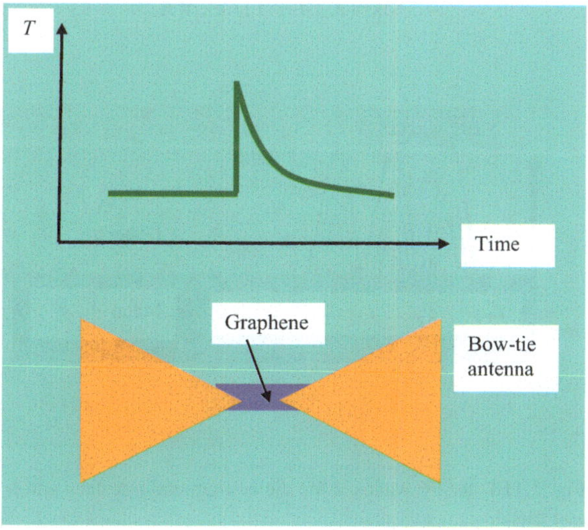

Fig. 1.117 The bolometric graphene detector

electron excitations generated by an electromagnetic field, which travel along a metal-dielectric interface, are localized around nanoparticles or nanowires placed on a substrate, or pass through an array of nanoholes. Because the wavelengths of SPP waves are smaller than the free space optical wavelength, miniaturization and scalability of optical devices are possible, although little progress was done in this respect up to now. However, SPP remains one of the most sensitive sensors for many analytes.

The simplest SPP configuration is the metal-dielectric interface, represented in Fig. 1.118. The SPP oscillations are localized in the z-direction, perpendicular to the interface. SPPs are electromagnetic fields with transverse magnetic (TM) polarization propagating in the x direction with a wavevector component k_{SPP}, which exist only if

$$\varepsilon(\omega)K = -\varepsilon_r K_m \tag{1.128}$$

where $\varepsilon(\omega)$ is the dielectric function of the metal and ε_r that of the dielectric, and $K_m = \sqrt{k_{SPP}^2 - \varepsilon(\omega)\omega^2/c^2}$ and $K = \sqrt{k_{SPP}^2 - \varepsilon_r\omega^2/c^2}$ are the wavevector components along z in metal and dielectric, respectively. The SPP theory and its applications are reviewed in Dragoman and Dragoman (2008b).

The advantages of SPP in graphene are that the SPP resonant frequencies, which occur generally at optical frequencies at metal/dielectric interfaces, could shift at much lower values, in the THz range, by applying gate voltages and that SPP waves in graphene display stronger electromagnetic confinement and lower dissipation losses compared to SPP waves in metallic structures. To illustrate the effect of graphene on the dispersion relation of SPPs, let's consider a thin metal slab of thickness d, in the limit $d \to 0$. SPP are TM waves at the metal-dielectric interfaces, but at very small metal thicknesses the SPP waves propagating at the two metal-dielectric interfaces couple, so that even and odd TM modes are produced, only the odd TM modes propagating with the dispersion relation (Jablan et al. 2013)

$$\exp(K_m d + 1)/\exp(K_m d - 1) = -\varepsilon(\omega)K/\varepsilon_r K_m \tag{1.129}$$

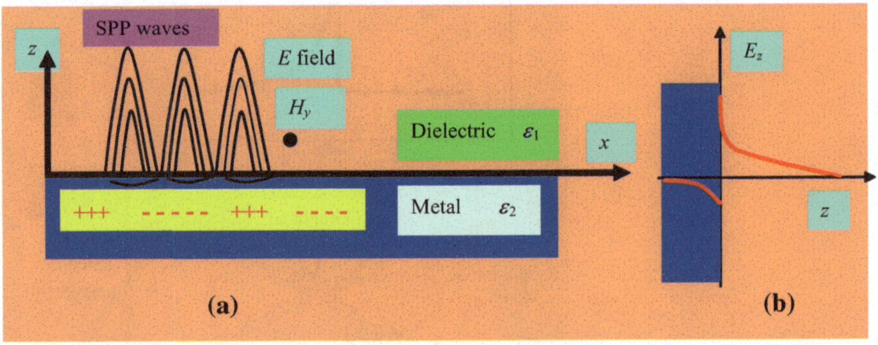

Fig. 1.118 **a** SPP waves at the dielectric/metal interface and **b** field penetration depth in dielectric and metal

The dielectric function of the bulk metal can be expressed as $\varepsilon(\omega) = 1 + i\sigma_v(\omega)/(\omega\varepsilon_0)$, where $\sigma_v(\omega)$ is the volume conductivity of the metal. In a very thin metal slab, however, the relative parameter is the surface conductivity defined as $\sigma_s = \sigma_v d$ and, in the limit $K_m d \ll 1$, we obtain the dispersion relation $K = 2i\omega\varepsilon_0\varepsilon_r/\sigma_s(\omega)$. In graphene, $\sigma_s(\omega)$ must be replaced by the graphene conductivity, $\sigma_G(\omega)$, which depends on the DC gate voltage, so that both SPP dispersion loss and resonant frequency are tunable.

If the volume conductivity is expressed by the Drude model:

$$\sigma_v = (in_v e^2/m_{eff})(\omega + i/\tau)^{-1} \tag{1.130}$$

where τ is the relaxation time and the volume electron density n_v is related to the surface densitity of carries n_s as $n_s = n_v d$, in the limit $\tau \to \infty$ we obtain $\varepsilon(\omega) = 1 - \omega_p^2/\omega^2$, with $\omega_p = n_v e^2/\varepsilon_0 m_{eff}$ the oscillation frequency of the volume plasma. For a thin metal slab the dispersion relation in the subwavelength limit, i.e. in the approximation $k_{SPP} \gg \omega/c$, we obtain

$$k_{SPP} = (2m_{eff}\varepsilon_0\varepsilon_r/n_s e^2)\omega(\omega + i/\tau) \tag{1.131}$$

In the generalization of the Drude model for graphene valid for large graphene doping, which corresponds to small SPP energies, the surface conductivity becomes

$$\sigma_D(\omega) = (ie^2 v_F n_s^{1/2}/\hbar\pi^{1/2})(\omega + i/\tau)^{-1} \tag{1.132}$$

where $n_s = \alpha V_G$. In this case, the gate-voltage-dependent SPP dispersion relation in graphene is given by

$$k_{SPP} = (2\hbar\varepsilon_0\varepsilon_r\pi^{1/2}/v_F n_s^{1/2} e^2)\omega(\omega + i/\tau) \tag{1.133}$$

For large SPP energies, we must take into account the interband transitions, case in which

$$\sigma_I = \frac{e^2}{4\hbar}\left(\theta(\hbar\omega - 2E_F) - \frac{i}{\pi}\ln\left|\frac{2E_F + \hbar\omega}{2E_F - \hbar\omega}\right|\right) \tag{1.134}$$

and the dispersion relation depends also on the Fermi level/doping in graphene. The total graphene conductivity is $\sigma_s = \sigma_D + \sigma_I$.

For frequencies below $2E_F$, σ_I is negative and there is no solution for SPP dispersion relation for TM waves. In this situation, a SPP dispersion relation for TE waves can be found as

$$K = \mu_0\omega i\sigma_s(\omega)/2. \tag{1.135}$$

Although light-plasmon coupling is much stronger in graphene than in conventional 2D semiconductors, the excitation of SPP modes in graphene is impeded by the mismatch between the wavenumber of free photons incident from air, where $k_{SPP} < \omega/c$, and SPP modes in the subwavelength region, for which $k_{SPP} \gg \omega/c$. A straightforward solution is to cut an array of ribbons in graphene, as suggested in Fig. 1.119. Then, plasmon waves in a ribbon with width W could be excited if their wavelength $\lambda_{SPP} = 2\pi/k_{SPP}$ satisfies the condition (Nikitin et al. 2012):

$$W \cong m\lambda_{SPP}/2 \tag{1.136}$$

where m is an integer. Equivalently, according to (1.133) in the weak damping limit $\tau \to \infty$, the light will be strongly absorbed/plasmon matching occurs for resonant frequencies

$$\omega_p \propto n_s^{-1/4} W^{-1/2} \propto V_G^{-1/4} W^{-1/2}, \tag{1.137}$$

which can be tuned by applied gate voltages V_G and/or by carefully choosing W. In particular, for a periodic arrays of graphene microribbons with widths chosen in the range 1–4 μm, and a gap:width ratio of 1, the SPP resonances are located in the THz region. For example, the SPP wavelengths can be tuned between 3 and 6 THz, as function of W, for a carrier density in graphene of $n_s = 1.5 \times 10^{13}$ cm^{-2}, or their

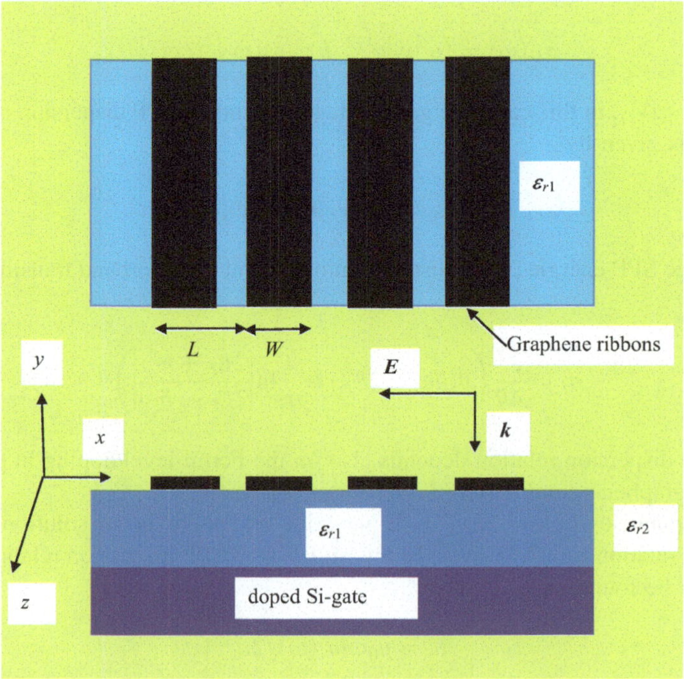

Fig. 1.119 Periodic array of graphene ribbons for SPP generation: top view (*top*) and side view (*bottom*)

frequencies vary between 3.6 and 6 THz for $W = 1$ μm if n_s changes between about 1.3×10^{13} cm^{-2} and 1.95×10^{13} cm^{-2} (Ju et al. 2011). The concentration of carriers n_s is simply modulated via gate voltages. The experimental structure needed to demonstrate the tuning of SPP frequencies via W or the carrier concentration is represented in Fig. 1.120, the top gate dielectric being ion gel.

Another way to excite SPPs is to use a silicon diffractive grating with a period Λ (Gao et al. 2013), as shown in Fig. 1.121. In this case, the dispersion relation can be written as

$$k_{SPP}(\omega) = \pi\hbar^2\varepsilon_0(\varepsilon_{r1} + \varepsilon_{r2})\omega^2(1 + i/\omega\tau)/e^2E_F \tag{1.138}$$

where E_F is the Fermi energy in graphene and ε_{r1} and ε_{r2} are the relative electric permittivities of the materials above and below graphene. Because the phase matching condition occurs when

$$\mathrm{Re}(k_{SPP}(\omega)) - \omega\sin\theta/c = 2\pi/\Lambda \tag{1.139}$$

where θ is the incidence angle of light with respect to the normal, under normal incidence, the SPP resonance frequency is given by

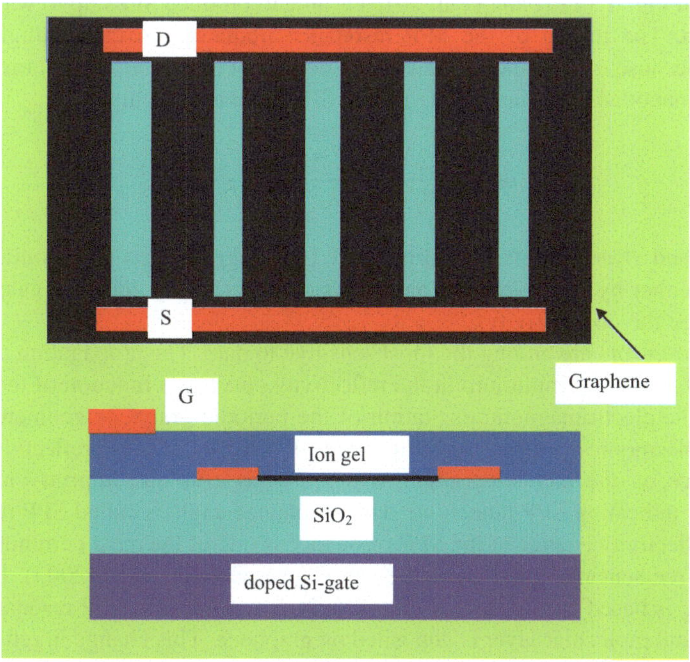

Fig. 1.120 Graphene micro-ribbons array for tunable THz SPP generation: *top view* before ion gel deposition (*top*) and side view (*bottom*)

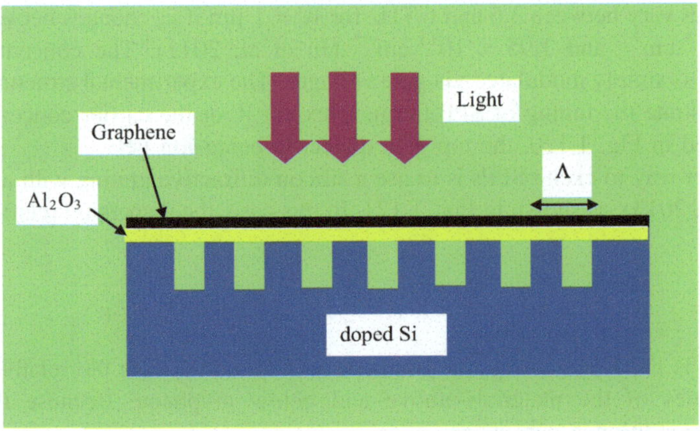

Fig. 1.121 SPP generation in graphene using a Si grating

$$\omega_{SPP} = \sqrt{2e^2 E_F / [\hbar^2 \varepsilon_0 (\varepsilon_{r1} + \varepsilon_{r2})\Lambda]} \tag{1.140}$$

and can be tuned via Λ or E_F.

Also, a periodic structure formed from graphene nanodisks or nanorings, as depicted in Fig. 1.122 (Fang et al. 2013) is able to produce SPPs up to wavelengths of 3.7 μm. The tuning of the SPP resonance frequency in this configuration is possible because of gate-induced electrical doping of graphene, the polarizability of graphene nanodisks of diameter D near a SPP resonance being

$$\alpha(\omega) = \frac{AD^3}{2B/(\varepsilon_{r1} + \varepsilon_{r2}) - i\omega D/\sigma(\omega)} \tag{1.141}$$

where A and B are constants. Formula (1.141) suggests that SPP tuning can be achieved either by modifying the nanodisks diameter, or by applying gate voltages that change the conductivity.

SPP biosensors are among the most sensitive to date. The propagation of SPPs is evidenced as a sharp minimum in the reflectivity curve as a function of the incident angle of the electromagnetic excitation of the periodic graphene geometry able to support plasmonics waves. As illustrated in Fig. 1.123, the reflectivity curve dependence on the incidence angle, $R = f(\varphi)$ has a unique imprint—a resonant behaviour indicating SPP launching. This resonance angle is called SPP resonance.

The reflectivity change at the SPP resonance is one of the most common concept for label-free sensing in biomedical sciences (Englebienne et al. 2003). Free-label biosensing is based on the shift of the reflectivity minimum at SPP resonance when an additional molecular layer is deposited on graphene. This change in reflectivity is extremely sensitive to additional molecules deposited on graphene, making SPP

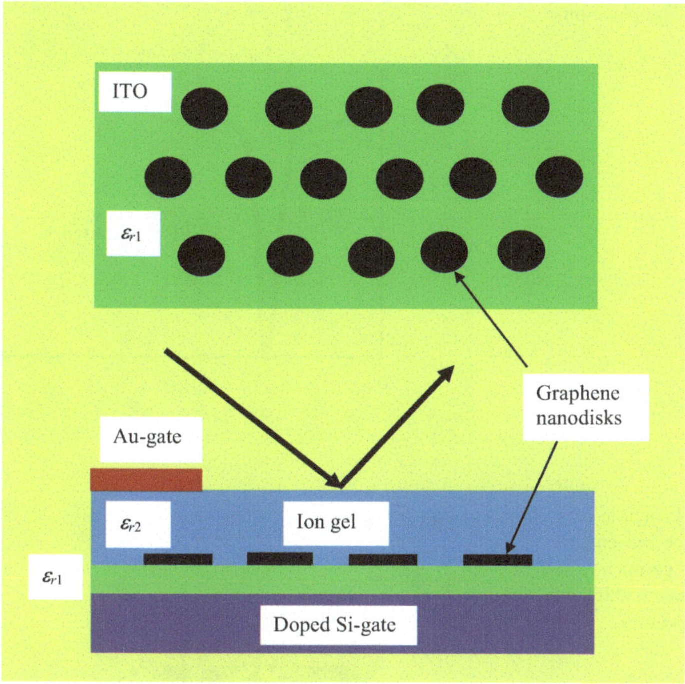

Fig. 1.122 SPP generation using a periodic graphene nanodisk array: *top view* before ion gel deposition (*top*) and side view (*bottom*)

resonance monitoring one of the most advanced methods to measure, for example, DNA hybridization.

The configuration of a label-free SPP biosensor is represented in Fig. 1.124. As the thickness of the sensing layer (graphene+biomolecules) increases because biomolecules are deposited on graphene, the incident angle of the optical excitation in the prism, φ_0, for which the reflectivity is minimum is modified by even few molecules.

The simple model of a three-layer (prism, metal+graphene sensing layer) sensing configuration can provide a quantitative description of the label-free SPP sensor. More precisely, variations in the SPP resonance angle are given by

$$\Delta\varphi = \alpha_1 \Delta n + \alpha_2 \Delta t \tag{1.142}$$

where α_1 and α_2 are constants and Δt represents the change in the thickness of the protein layer. The corresponding variation of the index of refraction is

$$\Delta n = -\frac{(n^2+2)^2}{6n}\frac{\Delta t}{t}\left(\frac{n^2-1}{n^2+2} - \frac{n_w^2-1}{n_w^2+2}\frac{V}{V_{tot}}\right) \tag{1.143}$$

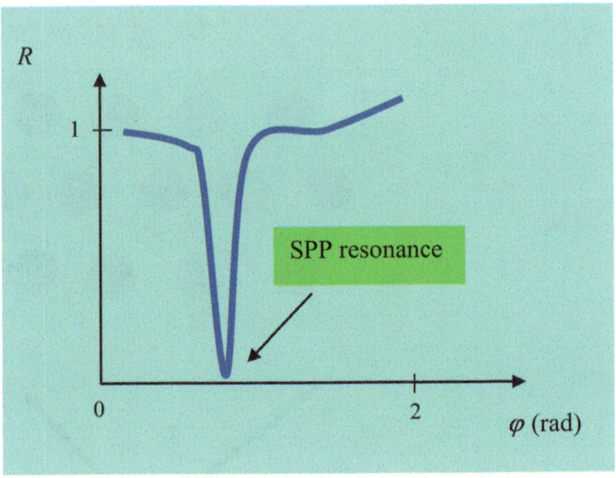

Fig. 1.123 The plasmonic resonance

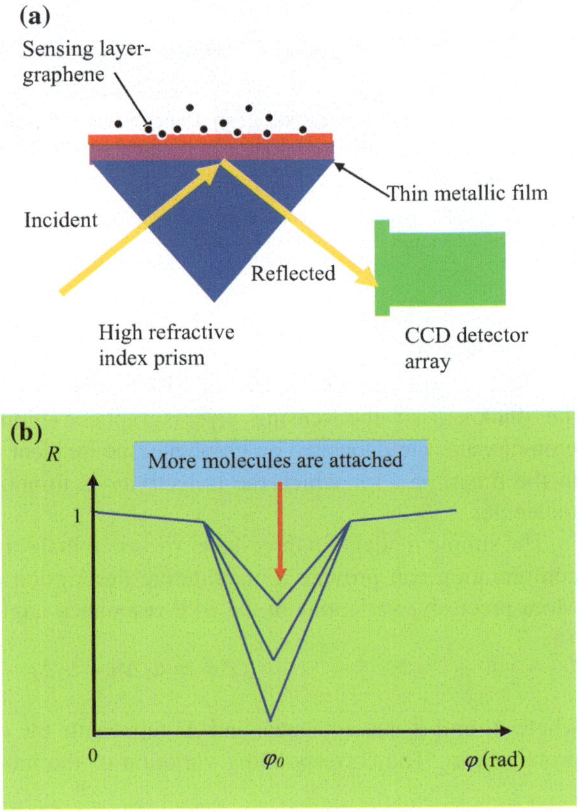

Fig. 1.124 Principle of a SPP label-free biosensor: **a** measuring geometry, and **b** SPP resonance shift with attached molecules

where n is the refractive index of the biomolecule with volume V and n_w is the refractive index of water, with volume V_w such that $V_{tot} = V + V_w$.

The changes of the SPP resonance angle are measured in resonance units (RU). 10^3 RU represents a modification with 0.1° of the SPP resonance angle due to biomolecules with a distributed mass of 1 ng/cm^2.

The sensitivity of SPP sensors is defined as the first derivative of the parameter (angle, wavelength) with respect to the parameter to be measured (refractive index, thickness, concentration, etc.). The sensitivity is measured in RIU (refractive index unit). The typical RIU is of the order 10^{-8}.

The SPP chip is in fact a gold film on glass or other dielectrics. In the case of SPP on graphene, the graphene monolayer acts as cover layer for the gold film. A graphene SPP biosensor having N graphene layers is g times more sensitive than SPP biosensors containing thin gold films (Wu et al. 2010a), where

$$g = (1 + 0.025\,N) \times \gamma \tag{1.144}$$

and $\gamma > 1$. The enhanced sensitivity is due to the optical properties of graphene and the increased adsorption of biomolecules. For $N = 10$, $\gamma = 4$ and the sensitivity increases with 25%.

The simplest method to produce a graphene SPP sensor on the graphene-metal surface is illustrated in Fig. 1.125. This graphene SPP chip obtained via transfer printing has many applications (Salihoglu et al. 2012). For instance, such graphene SPP chips are able to detect DNA hibridization with attomolecular concentration (Zagorodko et al. 2014). The conventional SPP instrumentation with nanomolar sensitivity is not sensitive enough to detect cancer markers or gene expressions, which require femto to attomolar level to detect DNA. To achieve such impressive sensitivities for DNA hybridization event detection, graphene, located on top of the SPP chip, must be noncovalently functionalized with gold nanostars carrying single stranded DNA. In this way, the SPP sensor attains a sensitivity of 500 aM for complementary DNA. The detection principle of DNA hybridization, shown in Fig. 1.126, is based on the significant change in the SPP resonance signal in the visible spectrum between non-hybridized and hybridized states of DNA.

Mid infrared is a special spectral region for biosensing because strong molecular vibrations of biomolecules (DNA, proteins, lipids) occur in this range. If this physical property is combined with the fact that SPPs in IR in graphene undergo an unprecedented spatial confinement, the result is a high sensitive SPP biosensor (Rodrigo et al. 2015). The biosensor consists of a graphene nanoribbon array (see Fig. 1.127) with $W = 30$ nm and $\Lambda = 80$ nm, placed on a Si/SiO$_2$ substrate, where the Si substrate is doped and serves as backgate. By changing the gate voltage between 0 and −100 V, the resonance frequency of the sensor changes from 1450 to about 1800 cm^{-1}. The functionality of this sensor is demonstrated by the detection of protein A/G, which binds to a goat anti-mouse IgG antibody to form a protein

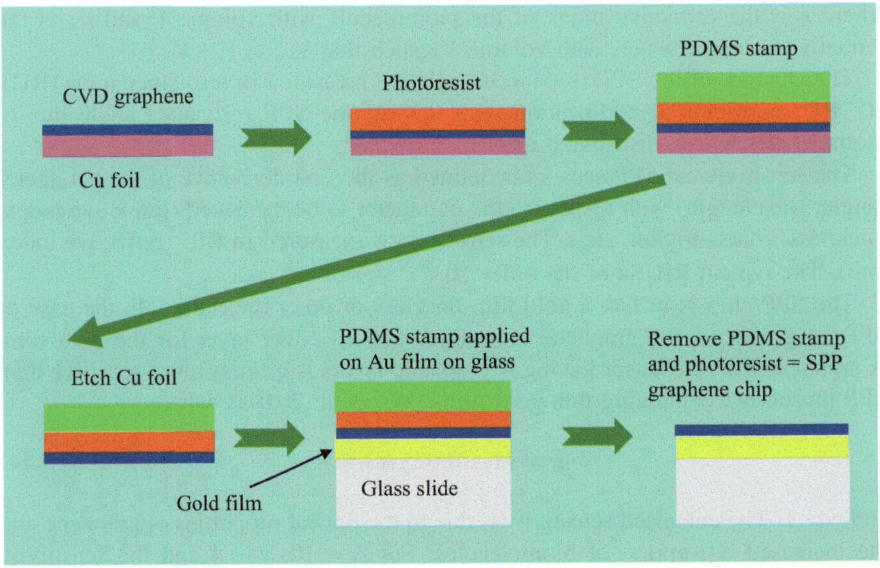

Fig. 1.125 SPP graphene chip fabrication

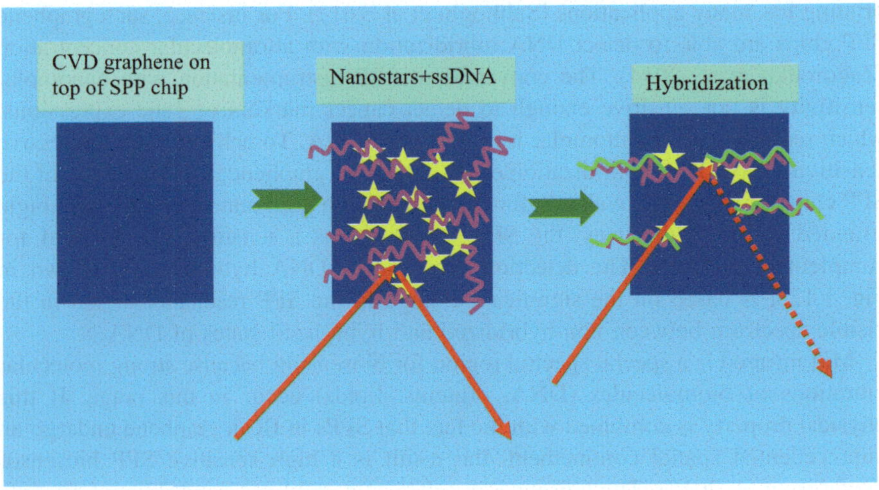

Fig. 1.126 Principle of DNA hybridization detection using a graphene chip

bilayer. The resonance shift produced by the resulting bilayer, which has a thickness of 8 nm, is 160 cm^{-1} in a graphene-based sensor, compared to 27 cm^{-1} when gold is used instead. From such measurements it is possible to extract the index of refraction of the protein in mid-IR.

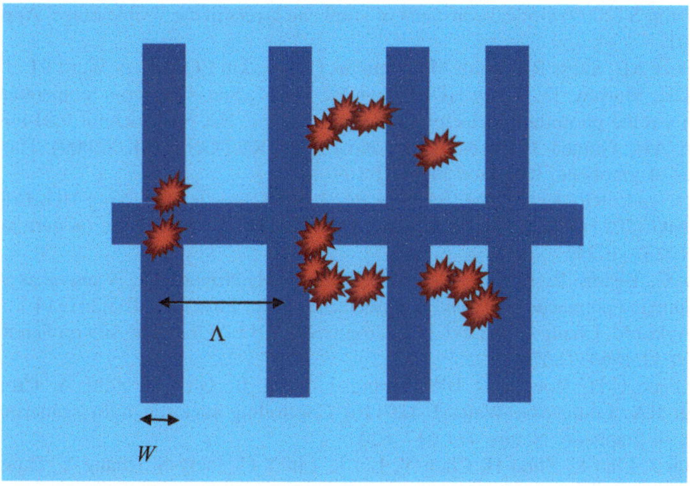

Fig. 1.127 Graphene nanoribbon array as mid IR biosensor

References

Bae S-H, Lee Y, Sharma BK, Lee HJ, Kim JH, Ahn J-H (2013) Graphene-based transparent strain sensor. Carbon 51:236–242

Balandin AA (2011) Thermal properties of graphene and nanostructured carbon materials. Nat Mater 10:569–581

Balocco C, Halsall M, Vinh NQ, Song AM (2008) THz operation of asymmetric-nanochannels devices. J Phys: Cond Matt 20:384203

Bao Q, Loh KP (2012) Graphene photonics, plasmonics, and broadband optoelectronic devices. ACS Nano 6:3677–3694

Baringhaus J, Ruan M, Edler F, Tejeda A, Sicot M, Taleb-Ibrahimi A, Li A-P, Jiang Z, Conrad EH, Berger C, Tegenkamp C, de Heer WA (2014) Exceptional ballistic transport in epitaxial graphene nanoribbons. Nature 506:349–354

Bekyarova E, Itkis ME, Ramesh P, Berger C, Spinkle M, de Heer WA, Handdon RC (2009) Chemical modification of epitaxial graphene: spontaneous grafting of aryl groups. J Am Chem Soc 131:1336–1337

Boland CS, Khan U, Backes C, O'Neill A, McCauley J, Duane S, Shanker R, Liu Y, Jurewicz I, Dalton AB, Coleman JN (2014) Sensitive, high-strain, high-rate bodily motion sensors based on graphene-rubber composites. ACS Nano 8:8819–8830

Bonaccorso F, Sun Z, Hasan T, Ferrari AC (2010) Graphene photonics and optoelectronics. Nat Photonics 4:611–622

Britnell L, Gorbachev RV, Jalil R, Belle BD, Schedin F, Mishchenko A, Georgiou T, Katsnelson MI, Eaves L, Morozov SV, Peres NMR, Leist J, Geim AK, Novoselov KS, Ponomarenko LA (2012) Field-effect tunneling transistor based on vertical graphene heterostructures. Science 335:947–950

Britnell L, Gorbachev RV, Geim AK, Ponomarenko LA, Mishchenko A, Greenaway MT, Fromhold TM, Novoselov KS, Eaves L (2013) Resonant tunneling and negative differential conductance in graphene transistors. Nature Commun 4:1794

Bolotin KI, Sikes KJ, Jiang Z, Klima M, Fudenberg G, Hone J, Kim P, Stormer HL (2008) Ultrahigh electron mobility in suspended graphene. Solid State Commun 146:351–355

Bruna M, Borini S (2009) Optical constants of graphene layers in the visible range. Appl Phys Lett 94:031901

Cai X, Sushkov AB, Suess RJ, Jadidi MM, Jenkins GJ, Nyakiti LO, Myers-Ward RL, Li S, Yan J, Gaskill DK, Murphy TE, Drew HD, Fuhrer MS (2014) Sensitive room-temperature terahertz detection via the photothermoelectric effect in graphene. Nat Nanotechnol 9:814–819

Castro Neto AH, Guinea F, Peres NMR, Novoselov KS, Geim AK (2009) The electronic properties of graphene. Rev Mod Phys 81:109–162

Cavallo F, Delgado RR, Kelly MM, Sánchez Pérez JR, Schroeder DP, Xing HG, Eriksson MA, Lagally MG (2014) Exceptional charge transport properties of graphene on germanium. ACS Nano 8:10237–10245

Chakraborty C, Beams R, Goodfellow KM, Wicks GM, Novotny L, Vamivakas AN (2014) Optical antenna enhanced graphene photodetector. Appl Phys Lett 105:241114

Chen C-C, Aykol M, Chang C-C, Levi AFJ, Cronin SB (2011a) Graphene-silicon Schottky diodes. Nano Lett 11:1863–1867

Chen C-F, Park C-H, Boudouris BW, Horng J, Geng B, Girit C, Zettl A, Crommie MF, Segalman RA, Louie SG, Wang F (2011b) Controlling inelastic light scattering quantum pathways in graphene. Nature 471:617–620

Cheng R, Bai J, Liao L, Zhou H, Chen Y, Liu L, Lin Y-C, Jiang S, Huang Y, Duan X (2012) High-frequency self-aligned graphene transistors with transferred gate stacks. Proc Nat Acad Sci 109:11588–11592

Deligeorgis G, Dragoman M, Neculoiu D, Dragoman D, Konstantinidis G, Cismaru A, Plana R (2010) Microwave switching of graphene field effect transistor at and far from the Dirac point. Appl Phys Lett 96:103105

Dong X, Shi Y, Huang W, Chen P, Li L-J (2010) Electrical detection of DNA hybridization with single-base specificity using transistors based on CVD-grown graphene sheets. Adv Mater 22:1–5

Dragoman D, Dragoman M (2007a) Negative differential resistance of electrons in graphene barrier. Appl Phys Lett 90:143111

Dragoman D, Dragoman M (2007b) Giant thermoelectric effect in graphene. Appl Phys Lett 91:203116

Dragoman D, Dragoman M (2008a) Terahertz Bloch oscillations in periodic graphene structures. Appl Phys Lett 93:103105

Dragoman M, Dragoman D (2008b) Plasmonics: Applications to nanoscale terahertz and optical devices. Prog Quantum Electron 32:1–41

Dragoman D (2009) Evidence against Klein paradox in graphene. Phys Scr 79:015003

Dragoman M, Dragoman D (2009) Graphene-based quantum electronics. Prog. Quantum Electron 33:165–214

Dragoman D (2010) Low-energy conductivity of single- and double-layer graphene from the uncertainty principle. Phys Scr 81:035702

Dragoman D, Dragoman M, Plana R (2010) Graphene-based ultrafast diode. Appl Phys Lett 108:084316

Dragoman D, Dragoman M (2012) Bionanoelectronics. Springer, Heidelberg

Dragoman M, Konstantinidis G, Tsagaraki K, Kostopoulos T, Dragoman D, Neculoiu D (2012a) Graphene-like metal-on-silicon field-effect transistor. Nanotechnology 23:305201

Dragoman M, Neculoiu D, Cismaru A, Deligeorgis G, Konstantinidis G, Dragoman D (2012b) Graphene radio: Detecting radiowaves with a single atom sheet. Appl Phys Lett 109:033109

Dragoman D, Dragoman M (2013) Geometrically induced rectification in two-dimensional ballistic nanodevices. J Phys D 46:055306

Dragoman D, Dragoman M (2014) Transport properties of carbon nanotubes and graphene. In: Tanaka K, Iijima S (eds) Carbon nanotubes and graphene, 2nd Edn. Elsevier, pp 151–164

Dragoman M, Dinescu A, Dragoman D (2014a) Negative differential resistance in graphene-based ballistic field-effect transistor with oblique top gate. Nanotechnology 25:41520

Dragoman M, Aldrigo M, Dinescu A, Dragoman D, Costanzo A (2014b) Towards a terahertz direct receiver based on graphene up to 10 THz. J Appl Phys 115:044307

Du X, Skachko I, Barker A, Andrei EY (2008) Approaching ballistic transport in suspended graphene. Nat Nanotechnol 3:491–495

Du X, Prober DE, Vora H, Mckitterick CB (2014) Graphene-based bolometers. Graphene 2D Mater 1:1–22

Englebienne P, Van Hoonacker A, Verhas M (2003) Surface plasmon resonance: principles, methods and applications in biomedical sciences. Spectroscopy 17:255–273

Fang T, Konar A, Xing H, Jena D (2007) Carrier statistics and quantum capacitance of graphene sheets and ribbons. Appl Phys Lett 91:092109

Fang Z, Thongrattanasiri S, Schlather A, Liu Z, Ma L, Wang Y, Ajayan PM, Nordlander P, Halas NJ, García de Abajo FJ (2013) Gated tunability and hybridization of localized plasmons in nanostructured graphene. ACS Nano 7:2388–2395

Fu X-W, Liao Z-M, Zhou J-X, Zhou Y-B, Wu H-C, Zhang R, Jing G, Xu J, Wu X, Guo W, Yu D (2011) Strain dependent resistance in chemical vapor deposition grown graphene. Appl Phys Lett 99:213107

Gao W, Shi G, Jin Z, Shu J, Zhang Q, Vajtai R, Ajayan PM, Kono J, Xu Q (2013) Excitation and active control of propagating surface plasmon polaritons in graphene. Nano Lett 13:3698–3702

Geim AK (2011) Nobel lecture: Random walk to graphene. Rev Mod Phys 83:851–862

Geim AK (2012) Graphene prehistory. Phys Scr T146:014003

Georgakilas V, Otyepka M, Bourlinos AB, Chandra V, Kim N, Kemp KC, Hobza P, Zboril R, Kim KS (2012) Functionalization of graphene: covalent and non-covalent approaches, derivatives and applications. Chem Rev 112:6156–6214

Green NS, Norton ML (2015) Interactions of DNA with graphene and sensing applications of graphene field-effect transistor devices: a review. Anal Chmica Acta 853:127–142

Guo Z, Dong R, Chakraborty PS, Lourenco N, Palmer J, Hu Y, Ruan M, Hankinson J, Kunc J, Cressler JD, Berger C, de Heer WA (2013) Record maximum oscillation frequency in C-face epitaxial graphene transistors. Nano Lett 13:942–947

Hähnlein B, Händel B, Pezoldt J, Töpfer H, Granzner R, Schwierz F (2012) Side-gate graphene field-effect transistors with high transconductance. Appl Phys Lett 101:093504

Han MY, Özyilmaz B, Zhang Y, Kim P (2007) Energy band-gap engineering of graphene nanoribbons. Phys Rev Lett 98:206805

Han S-J, Oida S, Jenkins KA, Lu D, Zhu Y (2013) Multifinger embedded T-shaped gate graphene RF transistors with high $f_{\max}/f_T$ ratio. IEEE Electron Dev Lett 34:1340–1342

Han S-J, Valdes Garcia A, Oida S, Jenkins KA, Haensch W (2014) Graphene radio frequency receiver integrated circuit. Nat Commun 5:3086

Hanson GW (2008) Dyadic Green's functions and guided surface waves for a surface conductivity model of graphene. J Appl Phys 103:064302

Hill EW, Vijayaragahvan A, Novoselov K (2011) Graphene sensors. IEEE Sens J 11:3161–3170

Hou C, Wang H, Zhang Q, Li Y, Zhu M (2014) Highly conductive, flexible, and compressible all-graphene passive electronic skin for sensing human touch. Adv Materials 26:5018–5024

Hu YT, Pantouvaki M, Brems S, Asselberghs I, Huyghebaert C, Geisler M, Alessandri C, Baets R, Absil P, Van Thourhout D, Van Campenhout J (2014) Broadband 10 Gb/s graphene electro-absorption modulator on silicon for chip-level optical interconnects. IEEE Electron Devices Meet (IEDM), 15–17 Dec 2014:5.6.1–5.6.4

Huang B-C, Zhang M, Wang Y, Woo J (2011a) Contact resistance in top-gated graphene field-effect transistors. Appl Phys Lett 99:032107

Huang M, Pascal TA, Kim H, Goddard III WA, Greer JR (2011b) Electronic-mechanical coupling in graphene from in situ nanoindentation experiments and multiscale atomistic simulations. Nano Lett:1241–1246

Jablan M, Soljačić M, Buljan H (2013) Plasmons in graphene: fundamental properties and potential applications. Proc IEEE 101:1689–1704

Ju L, Geng B, Horng J, Girit C, Martin M, Hao Z, Bechtel HA, Liang X, Zettl A, Shen YR, Wang F (2011) Graphene plasmonics for tunable terahertz metamaterials. Nat Nanotechnol 6:630–634

Jung M-H, Park G-H, Yoshida T, Fukidome H, Suemitsu T, Otsuji T, Suemitsu M (2013) High-performance graphene field-effect transistor with extremely small access length using self-aligned source and drain technique. Proc IEEE 101:1603–1608

Jurewicz I, Fahimi A, Lyons PE, Smith RJ, Cann M, Large ML, Tian M, Coleman JN, Dalton AB (2014) Insulator-conductor type transitions in graphene-modified silver nanowire networks: a route to inexpensive transparent conductors. Adv Funct Mat 24:7580–7587

Katsnelson MI (2012) Graphene. Carbon in Two-Dimensions. Cambridge University Press, Cambridge, UK:pp. 7–10

Katsnelson MI, Novoselov KS, Geim AK (2006) Chiral tunneling and the Klein paradox in graphene. Nat Phys 2:620–625

Kim KS, Zhao Y, Jang H, Lee SY, Kim JM, Kim KS, Ahn J-H, Kim P, Choi J-Y, Hong BH (2009) Large-scale pattern growth of graphene films for stretchable transparent electrodes. Nature 457:706–710

Kim BJ, Jang H, Lee S-K, Hong BH, Ahn J-H, Cho JH (2010) High-performance flexible graphene field effect transistors with ion gel gate dielectrics. Nano Lett 10:3464–3466

Kim D, Lee D, Lee Y, Jeon DY (2013a) Work-function engineering of graphene anode by bis (trifluoromethanesulphonyl)amide doping for efficient polymer light-emitting diodes. Adv Funct Mater 23:5049–5055

Kim H-Y, Lee K, McEvoy N, Yim C, Duesberg GS (2013b) Chemically modulated graphene diodes. Nano Lett 13:2182–2188

Kim U, Kang J, Lee C, Kwon HY, Hwang S, Moon H, Koo JC, Nam J-D, Hong BH, Choi J-B, Choi HR (2013c) A transparent and stretchable graphene-based actuator for tactile display. Nanotechnology 24:145501

Koenig SP, Doganov RA, Schmidt H, Castro Neto AH, Özyilmaz B (2014) Electric field effect in ultrathin black phosphorous. Appl Phys Lett 104:103106

Koppens FHL, Mueller T, Avouris Ph, Ferrari AC, Vitiello MS, Polini M (2014) Photodetectors based on graphene, other two-dimensional materials, and hybrid systems. Nat Nanotechnol 9:780–793

Kuila T, Bose S, Khanra P, Mishra AK, Kim NH, Lee JH (2011) Recent advances in graphene-based biosensors. Biosens Bioelectron 26:4637–4648

Kulkarni GS, Reddy K, Zhang Z, Fan X (2014) Graphene nanoelectronic heterodyne sensor for rapid and sensitive vapour detection. Nat Commun 5:4376

Kumar M, Bhaskaran H (2015) Ultrasensitive room-temperature piezoresistive transduction in graphene-based nanoelectromechanical systems. Nano Lett 15:2562–2567

Kumar S, Ahlawat W, Kumar R, Dilbaghi N (2015) Graphene, carbon nanotubes, zinc oxide and gold as elite nanomaterials for fabrication of biosensors for healthcare. Biosens Bioelectron 70:498–503

Kunc J, Hu Y, Palmer J, Guo Z, Hankinson J, Gamal SH, Berger C, de Heer WA (2014) Planar edge Schottky barrier-tunneling transistors using epitaxial graphene/SiC junctions. Nano Lett 14:5170–5175

Lee C, Wei X, Kysar JW, Hone J (2008) Measurement of the elastic properties and intrinsic strength of monolayer graphene. Science 321:385–388

Lee J, Ha T-J, Li H, Parrish KN, Holt M, Dodabalapur A, Ruoff RS, Akinwande D (2013) 25 GHz embedded-gate graphene transistors with high-k dielectrics on extremely flexible plastic sheets. ACS Nano 7:7744–7750

Lee J-H, Loya PE, Lou J, Thomas EL (2014) Dynamic mechanical behavior of multilayer graphene via supersonic projectile penetration. Science 346:1092–1096

Leong WS, Nai CT, Thong JTL (2014) What does annealing do to metal-graphene contacts? Nano Lett 14:3840–3847

Li J, Niu L, Zheng Z, Yan F (2014) Photosensitive graphene transistors. Adv Mater 26:5239–5273

Liao L, Bai J, Cheng R, Liu Y-C, Jiang S, Qu Y, Huang Y, Duang X (2010) Sub-100 nm channel length graphene transistors. Nano Lett 10:3952–3956

Liu M, Yin X, Ulin-Avila E, Geng B, Zentgraf T, Ju L, Wang F, Zhang X (2011) A graphene-based broadband optical modulator. Nature 474:64–67

Liu M, Yin X, Zhang X (2012) Double-layer graphene optical modulator. Nano Lett 12:1482–1485

Liu G, Ahsan S, Khitun AG, Lake RK, Balandin A (2013) Graphene-based non-Boolean logic circuits. J Appl Phys 114:154310

Liu J, Liu Z, Barrow CJ, Yang W (2015a) Molecularly engineered graphene surfaces for sensing applications: a review. Anal Chim Acta 859:1–19

Liu X, Sun J, Zhang X (2015b) Novel 3D graphene aerogel-ZnO composites as efficient detection for NO_2 at room temperature. Sens Actuators B 211:220–226

Low T, Avouris Ph (2014) Graphene plasmonics for terahertz to mid-infrared applications. ACS Nano 8:1086–1101

Lu C-C, Lin Y-C, Yeh C-H, Huang J-C, Chiu P-W (2012) High mobility flexible graphene field-effect transistors with self-healing gate dielectrics. ACS Nano 6:4469–4474

Luo S, Wang Y, Tong X, Wang Z (2015) Graphene-based optical modulators. Nanoscale Res Lett 10:199

Mayorov AS, Gorbachev RV, Morozov SV, Britnell L, Jalil R, Ponomarenko LA, Blake P, Novoselov KS, Watanabe K, Taniguchi T, Geim AK (2011) Micrometer-scale ballistic transport in encapsulated graphene at room temperature. Nano Lett 11:2396–2399

Mehr W, Dabrowski J, Scheytt JC, Lippert G, Xie Y-H, Lemme MC, Ostling M, Lupina G (2012) Vertical graphene base transistor. IEEE Electron Dev Lett 33:691–693

Merchant CA, Healy K, Wanunu M, Ray V, Peterman N, Bartel J, Fischbein MD, Venta K, Luo Z, Johnson ATC, Drndić M (2010) DNA translocation through graphene nanopores. Nano Lett 10:2915–2921

Miao X, Tongay S, Petterson MK, Berke K, Rinzler AG, Appleton BR, Hebard AF (2012) High efficiency graphene solar cells by chemical doping. Nano Lett 12:2745–2750

Mittendorff M, Winnerl S, Kamann J, Eroms J, Weiss D, Schneider H, Helm M (2013) Ultrafast graphene-based broadband THz detector. Appl Phys Lett 103:021113

Novoselov KS, Geim AK, Morozov SV, Jiang D, Zhang Y, Dubonos SV, Grigorieva IV, Firsov AA (2004) Electric field effect in atomically thin carbon films. Science 306:666–669

Novolesov KS, Geim AK, Morozov SV, Jiang D, Katsnelson MI, Grigorieva IV, Dubonos SV, Firsov AA (2005) Two-dimensional gas of massless Dirac fermions in graphene. Nature 438:197–200

Nikitin AYu, Guinea F, Garcia-Vidal FJ, Martin-Moreno L (2012) Surface-plasmon enhanced absorption and supressed transmission in periodic arrays of graphene ribbons. Phys Rev Lett B 85:081405

Park SJ, Kwon OS, Lee SH, Song HS, Park TH, Jang J (2012) Utrasensitive flexible graphene based field-effect transistor (FET)-type bioelectronic nose. Nano Lett 12:5082–5090

Park JJ, Hyun WJ, Mun SC, Park YT, Park OO (2015) Highly stretchable and wearable graphene strain sensors with controllable sensitivity for human motion monitoring. ACS Appl Mater Interfaces 7:6317–6324

Parrish KN, Akinwande D (2012) An exact solvable model for the graphene transistor in the quantum capacitance limit. Appl Phys Lett 101:053501

Petrone N, Meric I, Hone J, Shepard KL (2012) Graphene field-effect transistors with gigahertz-frequency power gain on flexible substrates. Nano Lett 12:121–125

Phare CT, Lee Y-HD, Cardenas J, Lipson M (2015) Graphene electro-optic modulator with 30 GHz bandwidth. Nat Photonics 9:511–515

Ponomarenko LA, Belle BD, Jalil R, Britnell L, Gorbachev RV, Geim AK, Novoselov KS, Castro Neto AH, Eaves L, Katsnelson MI (2013) Field-effect control of tunneling barrier height by exploiting graphene's low density of states. J Appl Phys 113:136502

Prezioso S, Perrozzi F, Giancaterini L, Cantalini C, Treossi E, Palermo V, Nardone M, Santucci S, Ottaviano L (2013) Graphene oxide as a practical solution to high sensitivity gas sensing. J Phys Chem C 117:10683–10690

Radoi A, Iordanescu A, Cismaru A, Dragoman M, Dragoman D (2010) Ultrabroadband photodetection based on graphene ink. Nanotechnology 21:455202

Reddy D, Register LF, Carpenter GD, Banerjee SK (2011) Graphene field-effect transistors. J Phys D 44:313001

Rodrigo D, Limaj O, Janner D, Etezadi D, Garcia de Abajo FJ, Pruneri V, Altug H (2015) Mid-infrared plasmonic biosensing with graphene. Science 349:165–169

Sabri SS, Levesque LP, Aguirre CM, Guillemette J, Martel R, Szkopek T (2009) Graphene field transistor with parylene gate dielectric. Appl Phys Lett 95:242104

Salihoglu O, Balci S, Kocabas C (2012) Plasmon-polaritons on graphene-metal surface and their use in biosensors. Appl Phys Let 100:213110

Schedin F, Geim AK, Morozov SM, Hill EW, Blake P, Katsnelson MI, Novoselov KS (2007) Detection of individual gas molecules adsorbed on graphene. Nat Mater 6:652–655

Schneider GF, Kowalczyk SW, Calado VE, Pandraud G, Zandbergen HW, Vandersypen LMK, Dekker C (2010) DNA translocation through nanopores. Nano Lett 10:3163–3167

Schwierz F (2013) Graphene transistors: status, prospects, and problems. Proc IEEE 101:1567–1584

Sensale-Rodríguez B, Yan R, Rafique S, Zhu M, Li W, Liang X, Gundlach D, Protasenko V, Kelly MM, Jena D, Liu L, Xing HG (2012) Extraordinary control of terahertz beam reflectance in graphene electro-absorption modulators. Nano Lett 12:4518–4522

Sensale-Rodríguez B, Yan R, Liu L, Jena D, Xiang HG (2013) Graphene for reconfigurable terahertz optoelectronics. Proc IEEE 101:1705–1716

Sevinçli H, Cuniberti G (2010) Enhanced thermoelectric figure of merit in edge disordered zigzag graphene nanoribbons. Phys Rev B 81:113401

Sharma BK, Ahn J-H (2013) Graphene based field effect transistors: efforts made towards flexible electronics. Solid State Electron 89:177–188

Sharma P, Bernard LS, Bazigos A, Magrez A, Ionescu AM (2015) Room-temperature negative differential resistance in graphene field effect transistors: Experiments and theory. ACS Nano 15:620–625

Sinha D, Lee JU (2014) Ideal graphene/silicon Schottky junction diodes. Nano Lett 14:4660–4664

Smith AD, Niklaus F, Paussa A, Vaziri S, Fischer AC, Sterner M, Forsberg F, Delin A, Esseni D, Palestri P, Östling M, Lemme MC (2013) Electromechanical piezoresistive sensing in suspended graphene membranes. Nano Lett 13:3237–3242

Song AM, Missous M, Omling P, Peaker AR, Samuelson L, Seifert W (2003) Unidirectional electron flow in a nanometer-scale semiconductor channel: a self-switching device. Appl Phys Lett 83:1881–1883

Stander N, Huard B, Goldhaber-Gordon D (2009) Evidence for Klein tunneling in graphene *p-n* junctions. Phys Rev Lett 102:026807

Sun Q, Kim DH, Park SS, Lee NY, Zhang Y, Lee JH, Cho K, Cho JH (2014) Transparent, low-power pressure sensor matrix based on coplanar-gate graphene transistors. Adv Mater 26:4735–4740

Sun Z, Chang H (2014) Graphene and graphene-like two-dimensional materials in photodetection: mechanisms and methodology. ACS Nano 8:4133–4156

Sutar S, Comfort ES, Liu J, Taniguchi T, Watanabe K, Lee JU (2012) Angle- dependent carrier transmission in graphene *p-n* junctions. Nano Lett 12:4460–4464

Tan Y-W, Zhang Y, Bolotin K, Zhao Y, Adam S, Hwang EH, Das Sarma S, Stormer HL, Kim P (2006) Measurement of scattering rate and minimum conductivity in graphene. Phys Rev Lett 99:246803

Thiele SA, Schaefer JA, Schwierz F (2010) Modeling of graphene metal-oxide-semiconductor field-effect transistors with gapless large-area graphene channels. J Appl Phys 107:094505

Tian H, Li C, Mohammad MA, Cui Y-L, Mi W-T, Yang Y, Xie D, Ren T-L (2014) Graphene earphones: entertainment for both humans and animals. ACS Nano 8:5883–5890

Tian H, Shu Y, Wang X-F, Mohammad MA, Bie Z, Xie Q-Y, Li C, Mi W-T, Yang Y, Ren T-L (2015) A graphene-based resistive pressure sensor with record-high sensitivity in a wide pressure range. Sci Rep 5:8603

Tongay S, Lamaitre M, Schumann T, Berke K, Appleton BR, Gila B, Hebard AF (2011) Graphene/GaN Schottky diodes: stability at elevated temperatures. Appl Phys Lett 99:102102

Tongay S, Lemaitre M, Miao X, Gila B, Appleton BR, Hebard AF (2012) Rectification at graphene-semiconductor interfaces: zero-gap semiconductor-based diodes. Phys Rev X 2:011002

Varghese SS, Lonkar S, Singh KK, Swaminathan S, Abdala A (2015) Recent advances in graphene based gas sensors. Sens Actuators B 218:160–183

Vicarelli L, Vitiello MS, Coquillat D, Lombardo A, Ferrari AC, Knap W, Polini M, Pellegrini V, Tredicucci A (2012) Graphene field-effect transistors as room-temperature terahertz detectors. Nat Mater 11:865–871

Vincenzi G, Deligeorgis G, Coccetti F, Dragoman M, Pierantoni L, Mencarelli D, Plana R (2012) Extending ballistic graphene FET lumped element models to diffusive devices. Solid-State Electron 76:8–12

Wang Y, Wong D, Shytov AV, Brar VW, Choi S, Wu Q, Tsai H-Z, Regan W, Zettl A, Kawakami RK, Louie SG, Levitov LS, Crommie MF (2013) Observing atomic collapse resonances in artificial nuclei on graphene. Science 340:734–737

Wei Z-J, Fu Y-Y, Liu J-B, Wang Z-D, Jia Y-H, Guo J, Ren L-M, Chen Y-F, Zhang H, Huang R, Zhang X (2014) Radio-frequency transistors from millimeter-scale graphene domains. Chin Phys B 23:117201

Wen Q-Y, Tian W, Mao Q, Chen Z, Liu W-W, Yang Q-H, Sanderson M, Zhang H-W (2014) Graphene based all-optical spatial terahertz modulator. Sci Rep 4:7409

Westlund A, Winters M, Ivanov IG, Hassan JU, Nilsson P-Å, Janzén E, Rorsman N, Grahn J (2015) Graphene self-switching diodes as zero-bias microwave detectors. Appl Phys Lett 106:093116

Whitener KE Jr, Sheehan PE (2014) Graphene synthesis. Diam Relat Mater 46:25–34

Wilmart Q, Berrada S, Torrin D, Nguyen VH, Fève G, Berroir J-M, Dollfus P, Plaçais B (2014) A Klein-tunneling transistor with ballistic graphene. 2D Mater 1:011006

Wu L, Chu HS, Koh S, Li EP (2010a) Highly sensitive graphene biosensors based on surface plasmon resonance. Opt Express 14:14395–14400

Wu YH, Yu T, Shen ZX (2010b) Two-dimensional carbon nanostructures: fundamental properties, synthesis, characterization and potential applications. J Appl Phys 108:071301

Xie B, Yang C, Zhang Z, Zou P, Lin Z, Shi G, Yang Q, Kang F, Wong C-P (2015) Shape-tailored graphene-based ultra-high-rate supercapacitor for wearable electronics. ACS Nano 9:5636–5645

Xu H, Zhang Z, Wang Z, Wang S, Liang X, Peng L-M (2011) Quantum capacitance limited vertical scaling of graphene field-effect transistor. ACS Nano 5:2340–2347

Xu H, Das S, Gong Y, Liu Q, Chien H-C, Chiu H-Y, Wu J, Hui R (2015) Complex refractive index tunability of graphene at 1550 nm wavelength. Appl Phys Lett 106:031109

Yasaei P, Kumar B, Hantehzadeh R, Kayyalha M, Baskin A, Repnin N, Wang C, Klie RF, Chen YP, Král P, Salehi-Khojin A (2014) Chemical sensing with switchable transport channels in graphene grain boundaries. Nat Commun 5:4911

Yao Y, Shankar R, Rauter P, Song Y, Kong J, Loncar M, Capasso F (2014) High-responsivity mid-infrared graphene detectors with antenna-enhanced photocarrier generation and collection. Nano Lett 14:3749–3754

Yavari F, Koratkar N (2012) Graphene-based chemical sensors. J Phys Chem Letter 3:1746–1753

Yeh C-H, Lain Y-W, Chiu Y-C, Liao C-H, Moyano DR, Hsu SSH, Chiu P-W (2014) Gigahertz flexible graphene transistors for microwave integrated circuits. ACS Nano 8:7663–7670

Young AF, Kim P (2009) Quantum interference and Klein tunneling in graphene heterojunctions. Nat Phys 5:222–226

Youngblood N, Anugrah Y, Ma R, Koester SJ, Li M (2014) Multifunctional graphene optical modular and photodetector integrated on silicon waveguides. Nano Lett 14:2741–2746

Zagorodko O, Spadavecchia J, Serrano AY, Larroulet I, Pesquera A, Zurutuza A, Boukherroub R, Szunerits S (2014) Highly sensitive detection of DNA hybridization on commercialized graphene-coated surface plasmon resonance interfaces. Anal Chem 86:11211–11216

Zak A, Andersson MA, Bauer M, Matukas J, Lisauskas A, Roskos HG, Stake J (2014) Antenna-integrated 0.6 THz FET direct detectors based on CVD graphene. Nano Lett 14:5834–5838

Zhang Y, Liu T, Meng B, Li X, Liang G, Hu X, Wang QJ (2013) Broadband high photoresponse from pure monolayer graphene photodetector. Nat Commun 4:1811

Zhao J, Zhang G-Y, Shi D-X (2013) Review of graphene-based strain sensors. Chin Phys B 522:057701

Zheng Q, Kim J-K (2015) Graphene for transparent conductors. Springer, New York

Zhu S-E, Ghatkesar MK, Zhang C, Janssen GCAM (2013a) Graphene based piezoresistive pressure sensor. Appl Phys Lett 102:161904

Zhu Z, Joshi S, Grover S, Moddel G (2013b) Graphene geometric diodes for terahertz rectennas. J Phys D 46:185101

Zubarev A, Dragoman D (2012) Tunable energy filtering of ballistic electrons in graphene. J Phys D 45:075301

Zurutuza A, Marinelli C (2014) Challenges and opportunities in graphene commercialization. Nat Nanotehnol 9:730–734

Zwolak M, Di Ventra M (2008) Physical approaches to DNA sequencing and detection. Rev Mod Phys 80:141–165

Chapter 2
Two-Dimensional Materials

Abstract In this chapter we present the recently discovered two-dimensional materials and their physical properties, which are useful for nanoelectronic devices. The synthesis methods play a central role in this chapter, since this is the key issue in the further development of this new emerging area of research. Details about characterization of these atomically thin materials are intrinsically linked to the growth methods.

2.1 Two-Dimensional Materials

The first two-dimensional (2D) material that was obtained is graphene, which is known for its unprecedented physical, mechanical, and thermal properties that often surpass similar properties of other materials. Graphene and its nanoelectronics applications were extensively studied in the first part of this book. Although references to graphene will appear also in the second part of the book from time to time, the main subject of this part is the other 2D materials, discovered in the last five years.

More than 140 2D materials, including graphene, are collected in the atlas of 2D materials in (Miró et al. 2014). The large majority of them originate in layered materials, held together by weak van der Waals forces of 40–70 meV (Butler et al. 2013; Kaul 2014). However, in a single atomically thin layer the atoms are strongly, covalently bonded to their neighbors. So, it is rather easy to exfoliate such single or few atomically thin layers and study them individually. Figure 2.1 represents schematically a layered material formed from several single atomically thin layers. There are a couple of families of 2D materials (Miró et al. 2014; Schwierz et al. 2015), the most known being:

X-enes which are atomically thin materials of a single element, with atoms arranged in a honeycomb lattice, e.g. graphene (carbon atoms), silicene (silicon atoms), germanene (germanium atoms), phosphorene (phosphorus atoms)

M. Dragoman and D. Dragoman, *2D Nanoelectronics*,
NanoScience and Technology, DOI 10.1007/978-3-319-48437-2_2

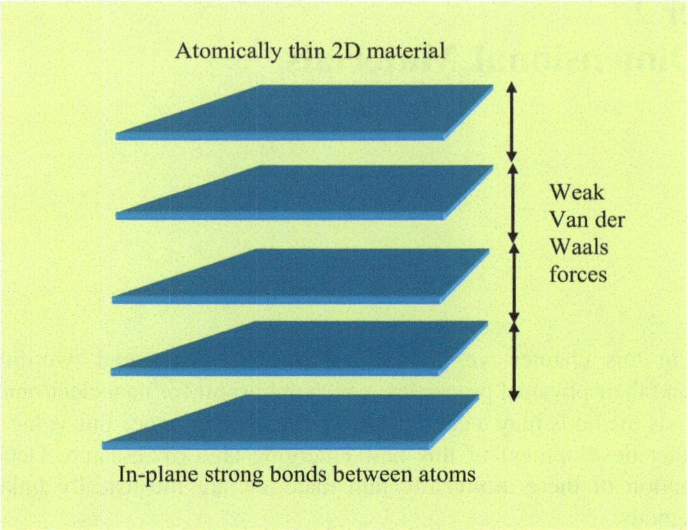

Fig. 2.1 Bulk material formed from several atomically thin monolayers

TMDs (transition metal dichalcogenides), of the form MX_2, where M stands for a transition metal (from the 4th, 5th, or 6th group of the periodic table of elements) and X is a chalcogen, such as S, Te, or Se. The TMDs based on M = Mo and W and others based on Hf, Pd, Pt and Zr are semiconductors. The most studied TMD semiconductors are MoS_2 and WS_2, which were the first atomically thin semiconductors

SMCs (semimetal chalcogenides), of the form M_2X_2, where M is a semimetal (Ga or In) and X a chalcogen (S or Se); they are semiconducting materials

MX-enes which have a hexagonal lattice and are of the form $M_{n+1}AX_n$, $n = 1, 2, 3$, where M is a transition metal, A is an element from group 13 or 14, and X carbon or hydrogen

Other 2D materials with hexagonal or tetragonal lattice structures could be produced, but some of them are unstable in air, whereas others could appear as a result of various experiments.

2D materials display many polytypes, such as 1T, 2H, 3R where T means trigonal, H signifies hexagonal and R stands for rhombohedral, while the number 1, 2, 3 shows the number of layers of the 2D materials in the unit cell. The main issue related to 2D materials is their growth. In this chapter we present the main growth methods and their present limitations (Butler et al. 2013; Chiu et al. 2015).

2.1.1 Bottom-up Growth Methods for TMDs

Mechanical exfoliation is based on mechanical cleavage of layered materials into one or few monolayers, possible due to weak van der Waals forces between adjacent layers. Mechanical exfoliation was used first to obtain graphene more than 10 years ago (Novoselov et al. 2004). Without this important achievement, this book and our work in the last 10 years in the area of 2D materials would have not existed. Now, the scotch tape method, named more scientifically mechanical exfoliation, is extended to all layered materials due to its simplicity. It is always accompanied by optical characterization, which determines where atomically thin materials are deposited on the substrate, and is followed by the indication of the respective area with alignment markers (see Fig. 2.2). Mechanical exfoliation is still a widespread method for producing high-quality 2D flakes. However, the yield of this method is low and finding the atomically thin materials is time consuming.

Liquid exfoliation via sonication produces large quantities of exfoliated flakes, which are either mono- or few-layer thick, and uses common organic solvents and/or functionalization. As an example, the Coleman method (Coleman et al. 2011) is based on sonication of TMDs powders in solvents with appropriately chosen surface tension, and is used to obtain nanosheets of different TMDs. Re-aggregation of nanosheets is avoided by using steric and electrostatic repulsion via adsorption control of molecules. The solvent, such as ethanol and water, is easily removed. Liquid exfoliation is a high-yield method and is improved by using aqueous surfactants and high-power sonication. This method is particularly suitable

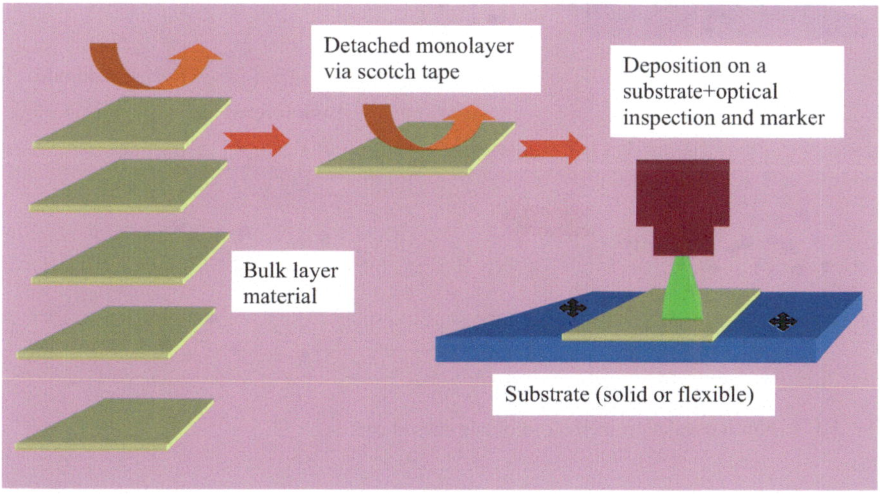

Fig. 2.2 Mechanical exfoliation method to obtain 2D materials

in various applications that require a large number of flakes, such as energy storage, catalysis and some applications in photonics.

Chemical lithium ion intercalation, represented in Fig. 2.3, is also a powerful method to obtain nanosheets. In this case, ions with small radii are intercalated into the gap between adjacent layers in layered materials. For instance, N-butyl lithium is commonly intercalated between monolayers of TMD materials. The gap is enlarged by the intercalated ions, and isolation of atomically thin flakes is achieved via gas formation following the reaction between the intercalated Li ions and water (see Chiu et al. (2015) and the references therein). The problem is that, during the intercalation of atoms, the 2H semiconductor MoS_2 transforms into its 1T metallic phase. However, the MoS_2 semiconducting phase is retrieved after annealing at 300 °C (Eda et al. 2011). Figure 2.3 illustrates the method of controllable intercalation of Li atoms for obtaining nanosheets using an electrochemical process (Zeng et al. 2011). The cathode is the TMD itself, and a Li foil is the anode. The intercalation of Li ions is controlled by a galvanic discharge process, which takes place at room temperature.

However, large flakes cannot be fabricated with any of the above-described methods. Instead, thinning techniques can be used to thin down a bulk MoS_2 crystal up to few layers or even monolayers. In this respect, lasers, thermal annealing and argon plasma are used for thinning.

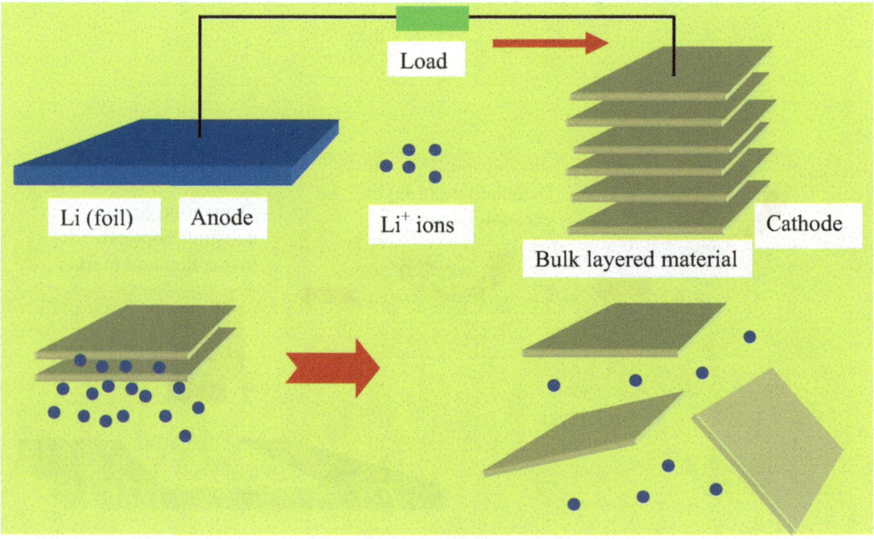

Fig. 2.3 Li ion intercalation method to obtain nanosheets

2.1.2 *Top-Down Growth Methods for TMDs*

The most important growth method of size-scalable TMDs is chemical vapor deposition (CVD). CVD is the best known method for growing graphene on 4-, 6- and even 12-in. Si/SiO_2 wafers. A simple CVD process is displayed in Fig. 2.4.

CVD growth of large-area MoS_2 monolayers was performed using as solid precursors MoO_3 and S powders. MoS_2 was grown on a Si/SiO_2 substrate at 650 °C and ambient pressure (Lee et al. 2012). The growth of MoS_2 is very sensitive to the state of the substrate. Therefore, before CVD growing, the substrate is treated with reduced graphene oxide, perylene-3,4,9,10-tetracarboxylic dianhydride (PTCDA) and perylene-3,4,9,10-tetracarboxylic acid tetrapotassium salt (PTAS).

Large area of MoS_2 can be grown directly on amorphous SiO_2. In this case, 0.4 g of MoO_3 powder is placed in a ceramic boat, as shown in Fig. 2.4, and the Si/SiO_2 substrate is oriented with the oxide side towards the MoO_3 boat. Another ceramic boat, containing 0.8 g sulfur powder, is placed in the reactor near the MoO_3 powder. The CVD takes place at 650 °C. Before growth, a droplet of reduced graphene oxide, PTCDA or PTAS in aqueous solution is spread on the substrate and then dried. As a result, MoO_{3-x} is formed at these elevated temperatures and diffuses into the substrate, further reactions with sulfur vapors generating MoS_2 in star shaped patches, which are linked to form a continuous film with a size of 2 mm and a thickness of 0.7 nm.

As for graphene, Raman analysis is used to identify MoS_2 monolayers. Two bands, at 403 and 385 cm^{-1}, corresponding to the A_{1g} and E_{2g} modes, with a full width half-maximum (FWHM) of 6.6 and 3.5 cm^{-1}, respectively, characterize this material. The difference between the peak frequencies of the A_{1g} and E_{2g} modes, denoted by Δ, determines the number of layers in the MoS_2 thin film. For values of

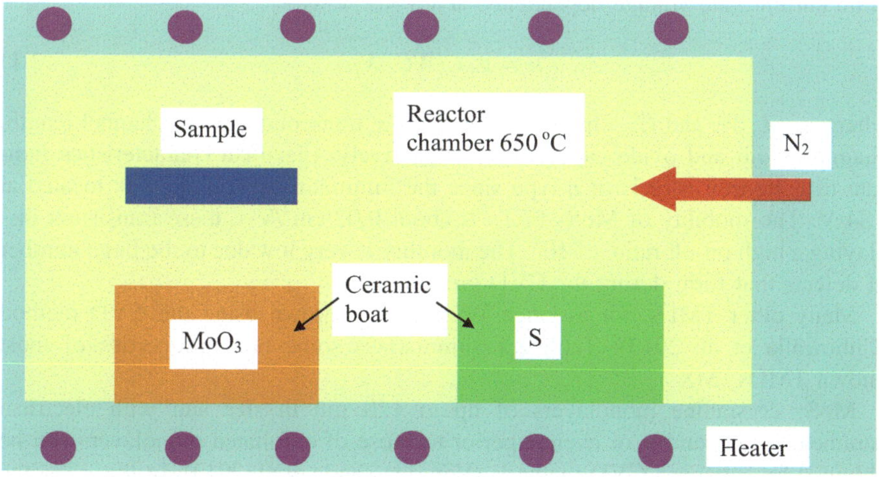

Fig. 2.4 CVD growth of MoS_2

Fig. 2.5 PL imprint of the MoS_2 monolayer

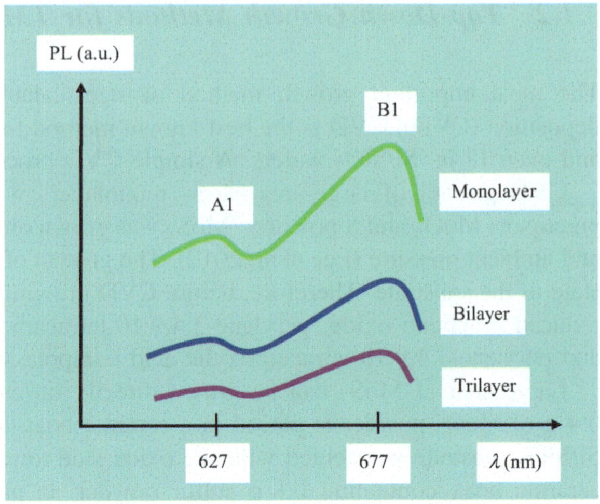

this parameter below 20 cm^{-1} we have MoS_2 monolayer, for Δ in the range of 20–21 cm^{-1} we obtain MoS_2 bilayer, while trilayer MoS_2 is formed if Δ is higher than 22 cm^{-1}. The photoluminescence (PL) spectrum displays two emission peaks, termed as A1 and B1, which correspond to exciton emissions located at 627 and 677 nm, respectively. The emission intensity decreases with the number of layers, indicating that the bandgap of MoS_2 changes from an indirect one in the case of bulk MoS_2 to a direct bandgap in the case of MoS_2 monolayer. This behavior is schematically represented in Fig. 2.5.

The DC characteristics are determined from measurements on a backgate FET fabricated via evaporating Au electrodes directly on the MoS_2 layer. The backgate transistor is depicted in Fig. 2.6. In this case, the mobility is extracted from the drain current-gate voltage dependence at V_D = 2 V as

$$\mu = g_m L / W C_{ox} V_D \tag{2.1}$$

where g_m, L, W, and C_{ox} signify, as before, the transconductance, channel length, channel width and oxide capacitance, respectively. Electrical characteristics indicate that the transistor is of n-type since the minimum drain voltage is located at −84 V. The mobility of MoS_2 FETs is about 0.02 cm^2/V s, these transistors displaying a high on-off ratio, of 10^4. The mobility is very low due to the large number of defects that form during the CVD growth.

Many other TMDs, for instance WS_2, can be grown using the CVD method (Chhowalla et al. 2013). Table 2.1 summarizes some basic properties of most known TMDs (MX_2).

MoS_2 crystalline monolayers of up to 120 μm in size and with electrical parameters comparable or even superior to those of exfoliated monolayers can be obtained by enhanced CVD methods (Van der Zande et al. 2013). In this case, the samples are grown from MoO_3 and S precursors, as above, without seeds for

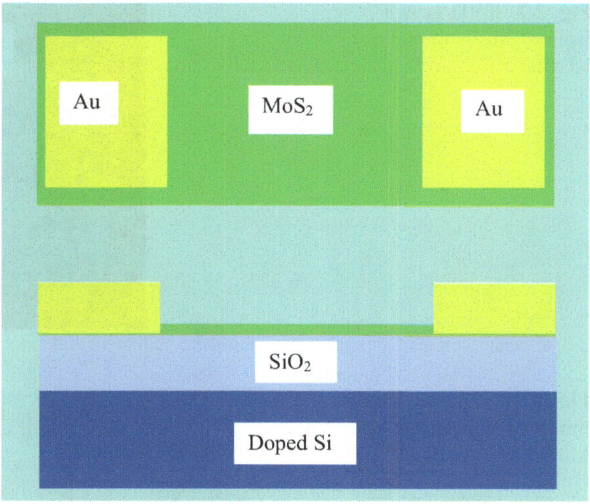

Fig. 2.6 Backgate transistor configuration for measuring the properties of the MoS_2 monolayer: top view (*top*) and side view (*bottom*)

Table 2.1 Some electrical properties of atomically thin semiconductor TMDs

M	X	Properties
Mo, W	S, Se, Te	Sulfides and selenides are semiconductors with a bandgap of 1 eV; telurides are semimetallic Electron and hole effective masses in $MoSe_2$: $m_e = 0.57\ m_0$, $m_h = -0.557\ m_0$ Electron and hole effective masses in WSe_2: $m_e = 0.33\ m_0$, $m_h = -0.46\ m_0$
Ti, Hf, Zr	S, Se, Te	All semiconductors, with bandgap in the range 0.2–2 eV
Pd, Pt	S, Se, Te	Sulfides and selenides are semiconductors with a bandgap of 0.4 eV, tellurides are metallic

growth nucleation. Isolated islands with triangular shapes and with edges in the range of 30–80 μm (typically) can be grown by this method (see Fig. 2.7). CVD MoS_2 monolayers are formed by connecting and overlapping such isolated islands. In particular, single-crystal triangles can aggregate in the form of two triangles intersecting at 40°–45°. Grain intersections are characterized by tilted and abrupt boundaries.

Transistors containing a single grain boundary can be fabricated. Multiple electrodes deposited on grain boundaries have the role of varying the channel length in backgate FET configurations. The FET electrodes can be positioned within the grain, along the grain boundary (parallel) and across the grain boundary (perpendicular). The multi-electrode FET is depicted in Fig. 2.8.

Various FETs, fabricated on similar mirror twin islands, are used to study the electrical properties of MoS_2 monolayers. The measured drain current-gate voltage dependence is a typical one, the extracted mobility is very low, of 1–8 cm^2/V s, and the on-off ratio at room temperature is of 10^5–10^7 cm^2/V s. However, since the

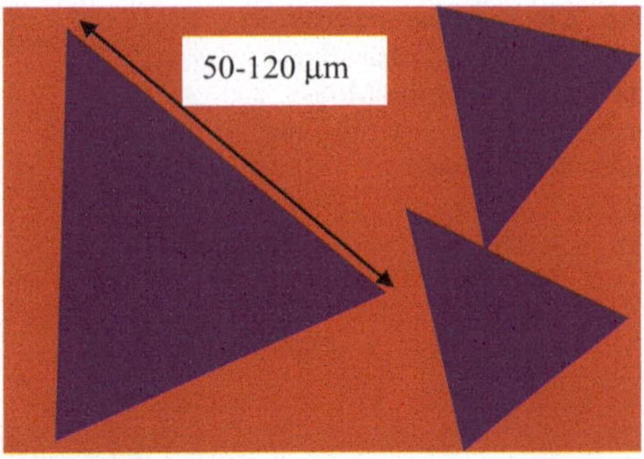

Fig. 2.7 CVD grown isolated MoS_2 monolayer islands

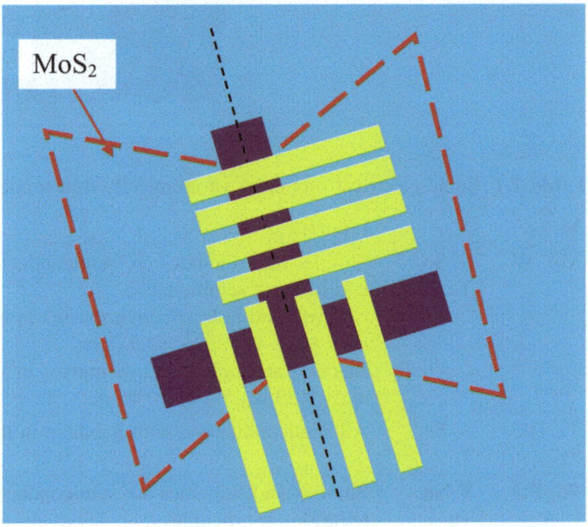

Fig. 2.8 FET on a single mirror twin grain boundary of MoS_2

dimensions of the transistors are higher than 10 μm, we are still far from the state-of-the-art nanoelectronics, were millions of transistors can be integrated on the same area. Monolayers have already dimensions of 120 μm, and the search continues to obtain larger monolayers, with better quality.

CVD can be performed with various precursors. For example, scalable and uniform MoS_2 monolayers and multilayers are grown via CVD at high temperatures, starting from $MoCl_5$ and sulfur (Yu et al. 2013). Monolayers with centimeter dimensions were grown on various substrates, such as SiO_2, graphite, or sapphire. The quality of MoS_2 layers were characterized as above, using Raman and PL, which are dependent on the layer thickness. Backgate FETs are used to extract the mobility, which is rather poor, of 0.003–0.03 cm^2/V s, and the drain currents, of up to 60 nA. Some precursors for TMD growth via CVD are displayed in Table 2.2 (Chhowalla 2013).

Despite the very good growing results, the devices are measured in selected areas and in reduced numbers, and the mobility is low, especially due to the

Table 2.2 CVD precursors

TMD	Precursor	Conditions
$MoSe_2$	$MoCl_5$ (255 °C), $C_4H_{10}Se$ (90 °C) or $C_8H_{18}Se$ (70 °C)	450–600 °C, atmosphere
MoS_2	$(NH_4)_2MoS_4$ in DMF	400 °C, 200 m Torr–2 Torr
WSe_2	$WOCl_6$ (260 °C), $C_4H_{10}Se$ (70 °C)	450–600 °C, atmosphere
WS_2	$WOCl_4$ (176 °C), $HS(CH_2)_2SH$ (63 °C)	250–600 °C, atmosphere

interaction between monolayers and the substrate. Therefore, the challenge is to grow TMDs at wafer scale and with higher mobilities.

Wafer-scale MoS_2 thin layers were obtained by sulfurization (Lin et al. 2012). In this case, thin films of MoO_3 were first deposited by thermal evaporation on a *c*-face sapphire substrate, and then placed in a quartz tube in the cold zone, where the Ar-H_2 mixture is flowing at a flow rate of 4:1 at 1 Torr (see Fig. 2.9). The furnace is heated at 500 °C and the MoO_3 on sapphire is placed in the center of the chamber, which is kept at low pressure for a first annealing. After one hour, a second annealing takes place in the presence of a sulfur boat at 1000 °C, the annealing rate being of 15 °C/min. Finally, the sample is cooled at room temperature.

The transfer of MoS_2 on the SiO_2 (300 nm)/Si substrate is performed by coating the MoS_2 thin film with PMMA via spin coating. The coated MoS_2 thin film was subsequently introduced in a NaOH solution at 100 °C for 30 min, and then in deionized water. The Si/SiO_2 substrate is pulling out the PMMA MoS_2 film from the solution. The PMMA is finally removed by acetone.

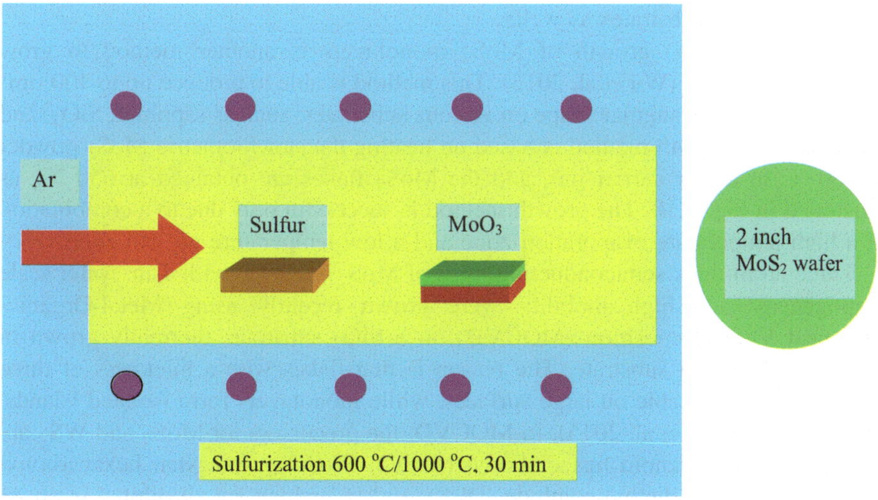

Fig. 2.9 Sulfurization method for MoS_2 growth at the wafer scale

This approach, which starts from a MoO_3 thin film, with a thickness of 3.6 nm, is scalable to a 2-in. sapphire wafer fabrication. The chemical reactions involved in this scalable method are

$$MoO_{3(s)} + H_{2(g)} \rightarrow MoO_{2(s)} + H_2O_{(g)}$$
$$MoO_{2(s)} + 2S_{(g)} \rightarrow MoS_{2(s)} + O_{2(g)}$$

the indices s and g indicating the phase (solid or gas) of the substances involved.

The thickness of the MoO_3 film is directly linked to the atomically thin MoS_2 layer, as follows:

- if MoO_3 has a thickness in the range of 0.8–2.2 nm, the MoS_2 thickness is between 1.52 nm (2 layers) and 2.3 nm (2–3 layers). However, the atomically thin layer morphology is formed from isolated islands
- if the thickness of MoO_3 is 3.6 nm, the thickness of MoS_2 is in the range 1.8–2 nm, i.e. 3 layers, and forms a continuous layer.

The electrical performances of MoS_2 are determined by measurements of a backgate transistor configuration. Since MoS_2 with a 3-atomic-layer thickness is already transferred on the wafer, Ti/Au electrodes are evaporated directly on MoS_2. The FET shows an *n*-type behavior and displays the following parameters:

- on-off ratio of 10^5
- mobility of 0.8 $cm^2/V\ s$

The channel is off at 0 V backgate voltage and reaches the maximum current of 1–5 μA at 180 V backgate voltage. These values do not differ from those of CVD grown MoS_2, but are obtained at the wafer level. The MoS_2 transfer can be performed on other substrates as well.

Vapor-solid (VS) growth of MoS_2 monolayers is another method to grow high-quality TMDs (Wu et al. 2013). This method is able to produce up to 400 μm^2 monolayers with triangular shape on various substrates, such as sapphire, SiO_2, and glass. The VS growth method is based on heating the stoichiometric MoS_2 powder at 900 °C in an Ar carrier gas, and the MoS_2 flakes are obtained at 650 °C, as illustrated in Fig. 2.10. The growth method is successful only due to a combination of a high-temperature evaporation zone and a low-temperature growth zone.

Three-atom-thick semiconducting films of MoS_2 and WS_2 with 4-in. wafer scale homogeneity and high mobility were grown recently using Metal-Organic-Chemical Vapor-Deposition (MOCVD) on a SiO_2 substrate, thermally grown in turn on a doped Si substrate. The reason is that flakes with a thickness of three atomic layers are stable on large surfaces, while monolayers form isolated islands, as shown in (Kang et al. 2015). In MOCVD, the precursors for MoS_2 and WS_2 are $Mo(CO)_6$—molybdenum hexacarbonyl (MHC), $W(CO)_6$—tungsten hexacarbonyl (THC), $(C_2H_5)_2S$—diethyl sulphide (DES), and H_2, which are all diluted in an Ar carrier gas. The concentration of each precursor is controlled during the entire growth process via pressure control. The process takes place in a 4-inch-diameter

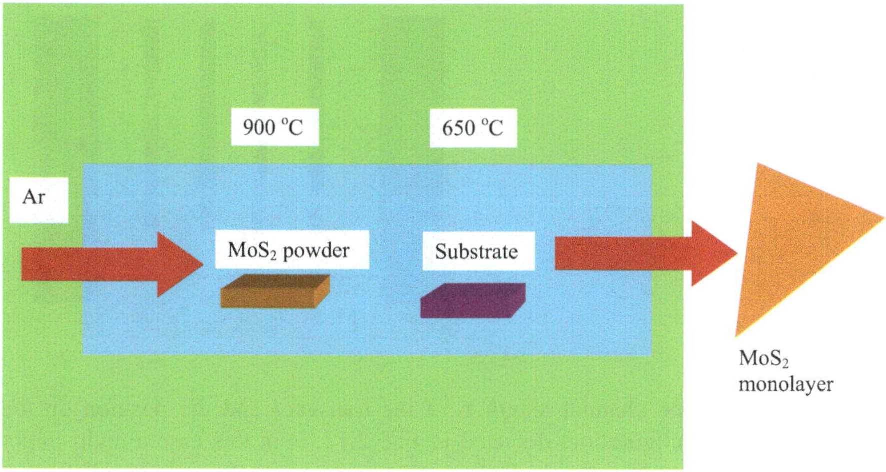

Fig. 2.10 VS growth method of triangular MoS_2 monolayers

hot-wall quartz tube, in which the precursors are injected as commercially available gases (see Fig. 2.11). The growth temperature is 550 °C and the pressure is maintained at 7 Torr during 26 h, which is a quite long time, necessary due to the low growth rate.

Based on this method, 8100 MoS_2 FETs in a backgate configuration originating from the common doped Si layer were fabricated on wafer with a yield around 99%. 100 transistors were measured to determine if the FET parameters, such as mobility,

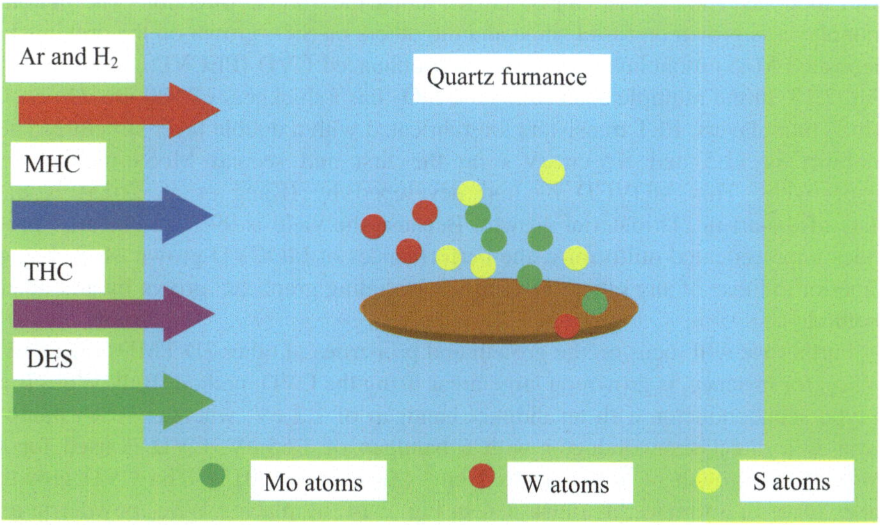

Fig. 2.11 MOCVD growth of three-atom-thick MoS_2 and WS_2 at the wafer level

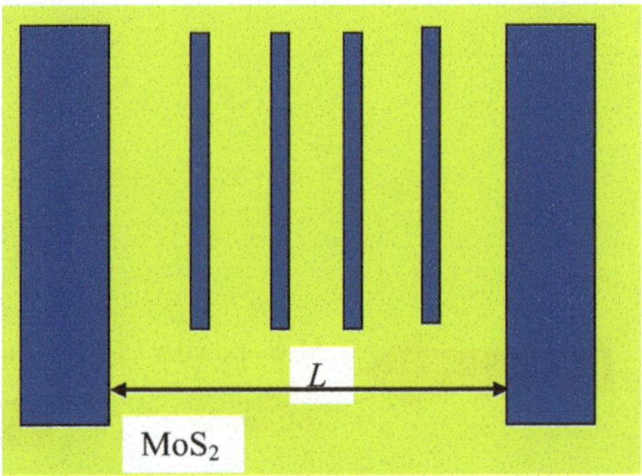

Fig. 2.12 MoS_2 FET backgate configuration on wafer

are dependent on the channel length L of the transistor and the position on the wafer. The FET configuration, depicted in Fig. 2.12, is in this case a multi-finger transistor, designed to measure the DC parameters at various drain-source distances. The maximum drain-source distance is 34 μm, and each two electrodes in Fig. 2.12 could play the role of source and drain, so that measurements could be performed for drain-source distances ranging from 1.6 up to 34 μm. Very small dispersion of FET parameters were observed.

Measurements of these FETs have confirmed a uniform and reversible conductance switching behavior from a high on state, of 10 μS, and an on-off ratio of 10^6. The carrier density is 7×10^{12} cm^{-2} for a backgate voltage of 5 V, and the mobility is 30 cm^2/V s. Gate-dependent sheet conductivity, mobility and other measurements show uniform results on different places on the wafer.

Multi-stacked MoS_2 monolayers can be produced with the MOCVD method for 3D electronic applications. In particular, three parallel substrates, each of which with monolayer thickness, can be grown along the vertical direction. The bottom monolayer is grown on fused silica and the others on SiO_2 grown on the previously deposited MoS_2 monolayers using plasma-enhanced CVD (PECVD), as shown in Fig. 2.13. In the multiple stacked layers, SiO_2 has a thickness of 100 nm. On each MoS_2 monolayers, FET transistors are fabricated with a double layer structure. The mobility is 11.5 and 8.5 cm^2/V s on the first and second MoS_2 monolayer, respectively. This MOVCD method developed by (Kang et al. 2015) is the state-of-the-art in 2D material growth. Because the yield is 99%, the method provides unprecedented uniformity, the performances of MOCVD-grown MoS_2 being superior to those of any other 2D material, including graphene, grown by any other method.

Further we will focus on the growth and properties of other 2D TMDs materials. WSe_2, for instance, is grown on large areas using the CVD method. Bulk WSe_2 is a *p*-type semiconductor with an indirect bandgap of 1.2 eV, whereas WSe_2 monolayer is a direct semiconductor with a bandgap of 1.65 eV. CVD is used for a controllable growth of WSe_2 up to 1 cm^2 (Zhou et al. 2015). The CVD growth takes place in a furnace, as illustrated in Fig. 2.14, by placing WSe_2 powder in an alumina boat at the center of the reactor. The SiO_2/doped Si substrate is used for

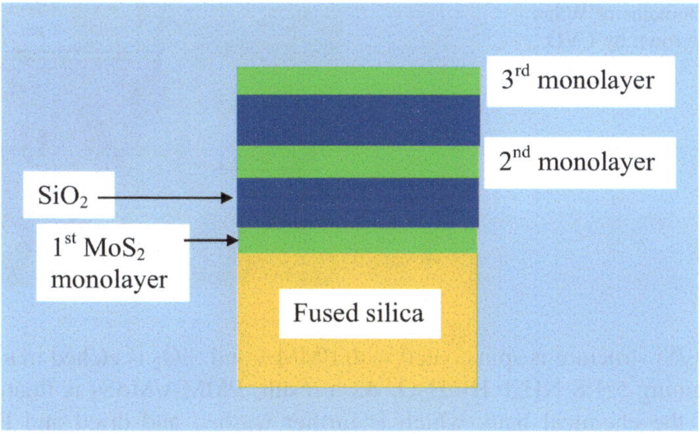

Fig. 2.13 Multiple layers of MoS_2 monolayer for 3D electronics

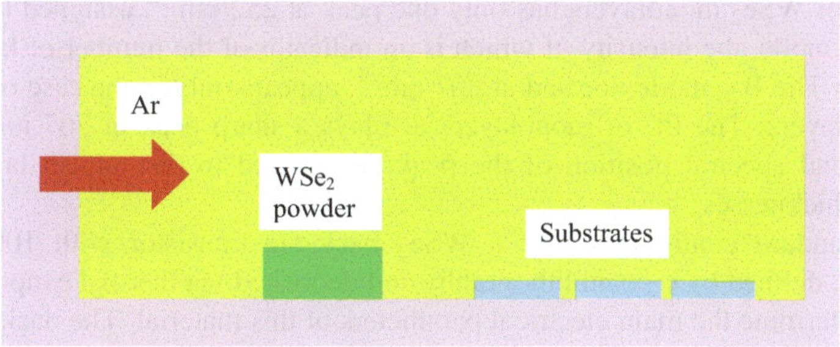

Fig. 2.14 CVD growth of WSe_2

WSe_2 atomic layer deposition. Argon gas is flowing in the reactor at a prescribed flow rate and the source temperature is constant and equal to 1060 °C. WSe_2 grows in triangular or hexagonal shapes with various dimensions, ranging from 10 to 50 μm (see Fig. 2.15).

At low substrate temperatures, of 750 °C, no visible WSe_2 domains form, but after annealing some nanosized domains are observed. At 765 °C, triangular domains appear, with 10 μm in size, having the nucleation density of 1060/mm^2. At higher temperatures, of 780 °C, larger domains with lower nucleation densities form. The optimum growth conditions for WSe_2 monolayers are reached at 765 °C and a flow rate of the carrier gas of 100 sccm. After 40 min under these growth conditions, the triangular and hexagonal domains link together and form a continuous monolayer. The WSe_2 surface, of 1 cm^2, could be increased if the dimensions of the furnace are increased (1-in. diameter quartz tube was used in the study of Zhou et al. (2015)). The transfer of WSe_2 monolayers onto an arbitrary substrate is performed by wet-etching of the SiO_2 layer. In this respect, the

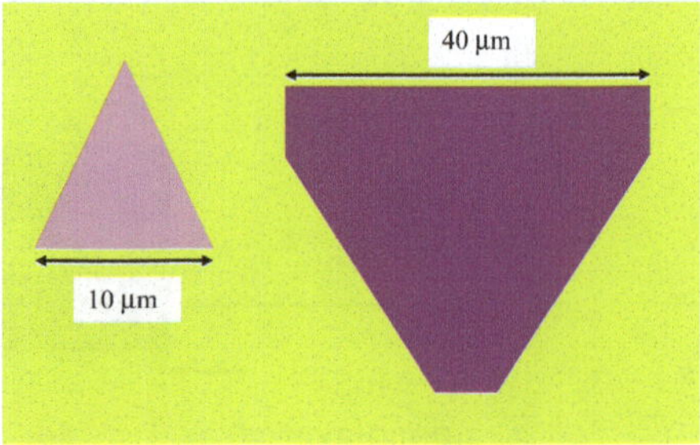

Fig. 2.15 Domains of WSe_2 monolayers grown by CVD

$WSe_2/SiO_2/Si$ structure is spin coated with PMMA and SiO_2 is etched in a chemical bath containing 5:1:8 NH_4F:HF:H_2O. As a result, PMMA/MoS_2 is floating at the surface of the chemical bath, which is further washed and dried and PMMA is removed using acetone.

The characterization of WSe_2 was done by a set of techniques, including Raman, PL, optical and electronic microscopy and electrical measurements. The Raman spectrum of WSe_2 monolayers has only one peak at 252 cm^{-1}, assigned to the A_{1g} resonance mode, the intensity of which is an indicator of the number of layers (see Fig. 2.16). The B_{2g} mode, located at 307 cm^{-1}, appears only in the case of bilayers and multilayers. The PL of monolayers displays a sharp peak at 767 nm, the PL intensity and spectral position of the peak being used to distinguish layers with different thicknesses.

The standard configuration of a WSe_2 backgate transistor with 100 nm Au electrodes, defined by e-beam lithography and deposited via thermal evaporation, is used to determine the main electrical parameters of this material. The backgate FET is fabricated on a monolayer domain, as depicted in Fig. 2.17. The current-voltage dependence indicates a *p*-type transistor with the highest mobility of 100 cm^2/V s and an on-off ratio of 10^8 when the backgate is swept between −100 and 100 V.

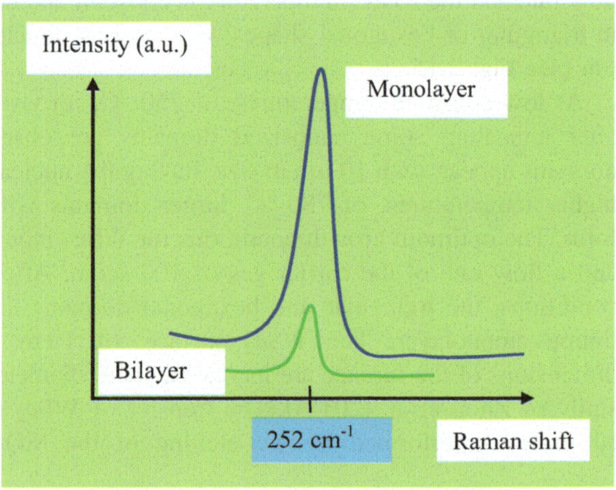

Fig. 2.16 Raman signature of WSe_2

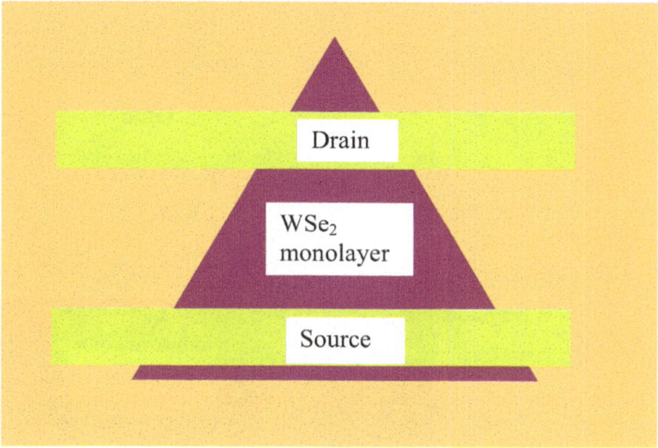

Fig. 2.17 Backgate FET with WSe_2 channel

The mobility varies in FETs with different WSe_2 thicknesses, the maximum mobility, of 350 cm^2/V s in a few-layer WSe_2 FET, being among the highest in TMDs.

A CVD-like method is also the selenization of WO_3 (Huang et al. 2014). WSe_2 monolayers can be grown via this method at 850 °C on a sapphire substrate, the obtained triangular flakes having sizes in the range of 10–50 μm. The method of selenization is based on the reaction between WO_3 and Se at high temperatures, as shown in Fig. 2.18. The chemical reaction is

$$WO_3 + 3Se + H_2 \rightarrow WSe_2 + H_2O + SeO_2$$

A similar method (sulfurization of MoO_3 at 650 °C) can be used to grow MoS_2 monolayers. In the Raman spectrum analysis, performed under excitation with a laser at 473 nm, two peaks appear for WSe_2 monolayers: one at 248 cm^{-1}, corresponding to the E^1_{2g} mode, and another at 259 cm^{-1}, associated to A_{1g}, as above.

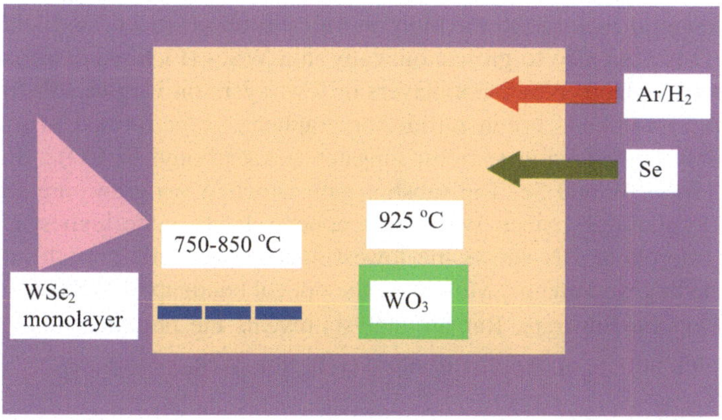

Fig. 2.18 Triangular WSe_2 monolayer obtained by selenization

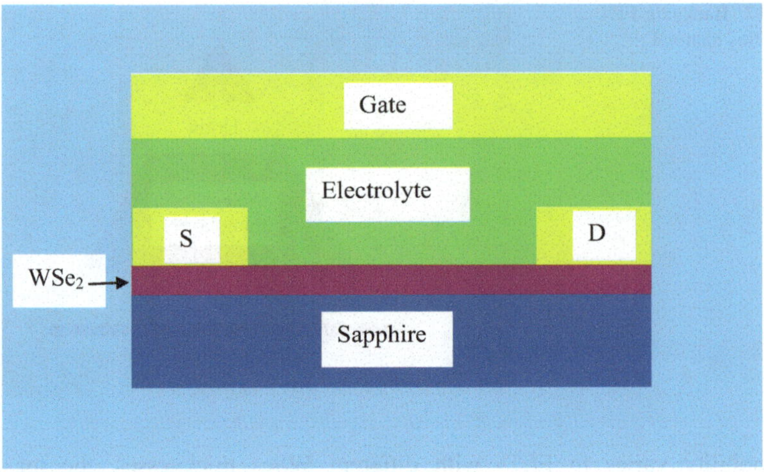

Fig. 2.19 Schematic representation of an EDLT

Again, the intensity and shift of Raman modes is used to discern the number of layers.

It is interesting to note that the evaluation of electrical properties of WSe_2 was performed with the help of an electric double-layer transistor (EDLT), able to determine the mobility at high concentrations of charge carriers. A double-layer transistor is represented in Fig. 2.19. The EDLT has a WSe_2 monolayer as channel, with a channel width of about 670 μm and a channel length of 1225 μm. The electrolyte is an ion gel, an electric double layer forming at the WSe_2/electrolyte interface. The drain current versus gate voltage dependence is measured, the extracted mobility being of 90 cm^2/V s for holes and of only 7 cm^2/V s for electrons. The on-off ratio is 10^5 for a *p*-channel EDLT and 10^4 for an *n*-channel transistor. The type of channel (*p* or *n*) is controlled by the polarization of the electric double layer on WSe_2 and is determined by the gate voltage. A very good inverter was produced due to the high on-off ratio observed in this EDLT.

MOCVD is used also to grow atomically thin WSe_2 (Eichfeld et al. 2015). The fabrication of scalable WSe_2 monolayers or few layers on various substrates, such as sapphire, amorphous boron nitride, or graphene, is performed in a cold-wall reactor using the following sources: tungsten hexacarbonyl $W(CO)_6$, dimethylselenium $(CH_3)Se$, W and Se. The substrates on which WSe_2 grows are selected for different morphologies, since WSe_2 on graphene displays the highest nucleation, WSe_2 on amorphous BN shows the lowest nucleation, while growth on sapphire provides the largest domains. Moreover, the optical bandgap of WSe_2 is tunable by the choice of the substrate. Raman analysis reveals the presence of E^1_{2g} and A_{1g} modes, which are the signatures of atomically thin WSe_2.

2.1.3 Growth of 2D TMDs Alloys and Heterostructures

The growth of the 2D alloy $MoS_{2(1-x)}Se_{2x}$ is performed by physical vapor deposition (PVD) (Feng et al. 2014b). In this method, $MoSe_2$ and MoS_2 powders are placed inside a furnace at two distinct zones, T1 and T2, with different temperatures, as shown in Fig. 2.20. Further, $MoSe_2$ and MoS_2 are vaporized at 940–975 °C. The Si substrate, on which a SiO_2 layer with a thickness of 270 nm is grown, is placed in another zone, T3, with a lower temperature, of 600–700 °C. A temperature gradient of 50°/cm is applied along the T3 zone in order to favor the growth of $MoS_{2(1-x)}Se_{2x}$ monolayers. Large areas of high-quality $MoS_{2(1-x)}Se_{2x}$ flakes are obtained by varying the growth conditions, such as evaporation temperature, hydrogen gas, and flow rate.

In this respect, the loading control of the two evaporation sources (powders of $MoSe_2$ and MoS_2) can tune the size and shape of monolayers. More precisely, in an alloy with $x = 0.20$, when small amounts are loaded, monolayers of triangular shape and sizes larger than 2 μm are produced, while by increasing the loading, continuous monolayers are fabricated as rectangles with an area of 0.5×0.5 cm^2. Raman and PL studies show that the peaks are shifted as a function of the x value in the alloy. In particular, the A exciton emission energy decreases from 1.86 up to 1.73 eV, corresponding to a redshift in wavelength from 665 to 716 nm, as the Se content increases from 0 to 0.4. Backgate FETs fabricated from alloys with different Se content, i.e. with $x = 0$, $x = 0.15$, and $x = 0.30$ (see Fig. 2.21), are of n-type, have high on-off ratios ($>10^5$) and a mobility of 0.1–0.4 cm^2/V s.

The method discussed above is suitable for growing $MoS_{2(1-x)}Se_{2x}$ monolayers with Se contents up to only 0.4, due to $MoSe_2$ decomposition into Mo_3Se_4 at high temperatures. For higher values of x, ranging from 0.41 up to 1, the same group of authors used controlled morphology in PVD with added Se vapor (Feng et al. 2015). The PVD process is displayed in Fig. 2.22. The Se powder is placed in the

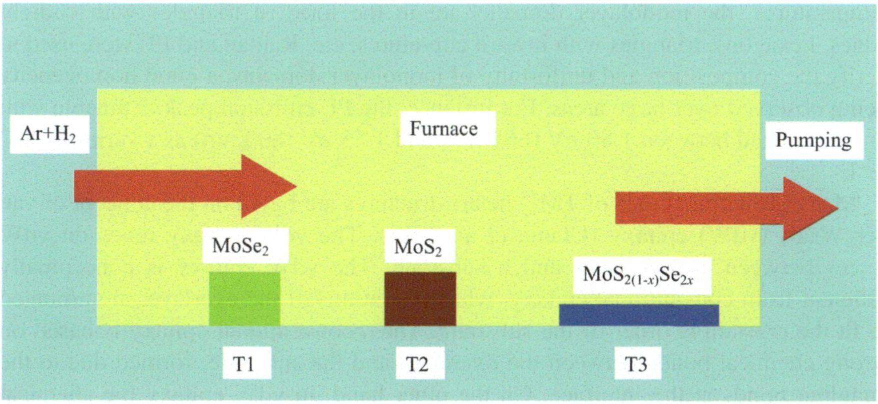

Fig. 2.20 PVD growth of $MoS_{2(1-x)}Se_{2x}$

Fig. 2.21 $MoS_{2(1-x)}Se_{2x}$ FET configuration

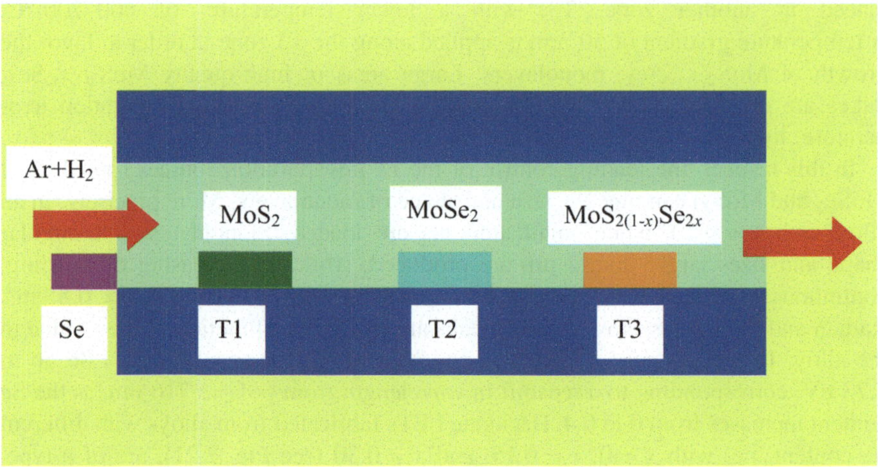

Fig. 2.22 PVD growth of $MoS_{2(1-x)}Se_{2x}$ with x in the range 0.41–1

gas stream, while the MoS_2 and $MoSe_2$ powders are arranged in two different temperature zones of the furnace and vaporized at 950–965 °C. The SiO_2/Si substrate is placed in the area with a lower temperature (600–770 °C). Monolayers with sizes up to 20 μm are obtained by controlling growth parameters, such as evaporation temperature, temperature gradient, etc. Depending on the different deposition temperatures, the monolayer domains are in the form of triangles with straight edges, hexagons, triangles with inward curvatures, etc. Raman and PL were used to verify the composition and uniformity of monolayer domains, a good homogeneity being observed over large areas. For instance, the PL emission peak is tunable with x: it is situated between 1.86 eV (665 nm) and 1.55 eV (800 nm) as x varies from 0 to 1.

Many growth methods of TMD heterostructures are based on the concept of van der Waals (vdW) epitaxy (Utama et al. 2013). The vdW epitaxy relies on vdW forces between an overlayer and a substrate. The vdW epitaxy is conceptually different from conventional epitaxy, where the atoms of the overlayer are arranged to fit the crystalline order of the substrate. Thus, conventional epitaxy is based on strong chemical bonds between the overlayer and the substrate, formed due to the dangling bonds at the interface. On the other hand, in vdW epitaxy the chemical bonds that keep atoms together do not form between monolayers in the staking, but

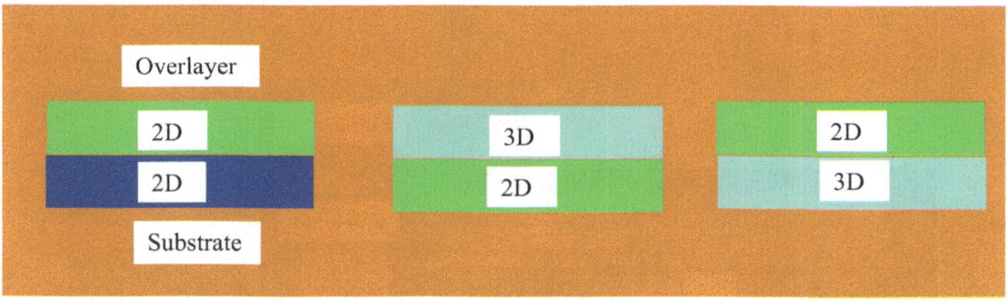

Fig. 2.23 The vdW epitaxy principles

are present eventually only in the in-plane monolayers. Multilayered materials, such as TMDs, have no dangling-bonds, since the monolayers are connected only by vdW forces. In principle, in a vdW growth either the substrate or the overlayer must be free of dangling bonds, and this allows the growth of heterostructures with a very large lattice mismatch (around 40%) in several ways (see Fig. 2.23). The vdW epitaxy is known since the 70s, was extensively used 20 years ago (Utama et al. 2013), then abandoned and rediscovered in the context of 2D materials (Geim and Grigorieva 2013). Graphene has a similar history, being observed first 50 years ago, but never studied in-depth until 2004, by the Geim group.

A first example of the vdW concept is the heterostructure involving WS_2 and MoS_2 monolayers, where both 2D materials were grown by CVD (Tongay et al. 2014), and WS_2 is the top material transferred on MoS_2 by the PDMS stamping method. The vertical heterostructure, with an area of 20×40 μm^2, uses SiO_2/Si as substrate. In this case, Raman or PL analysis show peaks corresponding to the two types of monolayers, each monolayer contributing independently to the total Raman or PL signature of the entire heterostructure. This situation is not encountered in any type of heterostructure where interaction/bandgap renormalization occurs. The explanation of the lack of interaction between monolayers in the heterostructure, valid also for WS_2/MoS_2 monolayer heterostructures, is the large interlayer separation. In the case of WS_2/MoS_2 heterostructure, the interlayer separation is due to residues trapped between monolayers during the transfer process. It should be noted that the interlayer distance is reduced by annealing at 120 °C, for 6 h, in vacuum, operation which has as final result the change of PL and Raman spectra.

All-CVD growth of MoS_2/h-BN vertical vdW heterostructures was performed by (Wang et al. 2015). This is still a hybrid method, as above, because the MoS_2 and h-BN monolayers in the heterostructure are not systematically grown one on top of the other, as in the case of molecular beam epitaxy (MBE) growth. On the contrary, in the all-CVD method, the h-BN layer is first grown by CVD on copper foil and then transferred to a SiO_2 chip using PMMA-assisted techniques. The chip is introduced in a furnace, where the MoS_2 monolayer is grown over it, covering the entire chip. This growth method is depicted in Fig. 2.24. The precursors, MoO_3 and S, are placed in another furnace, different from that containing the h-BN/SiO_2 chip, in order to have a better control of the temperature. In Fig. 2.24, MoO_3 is located in

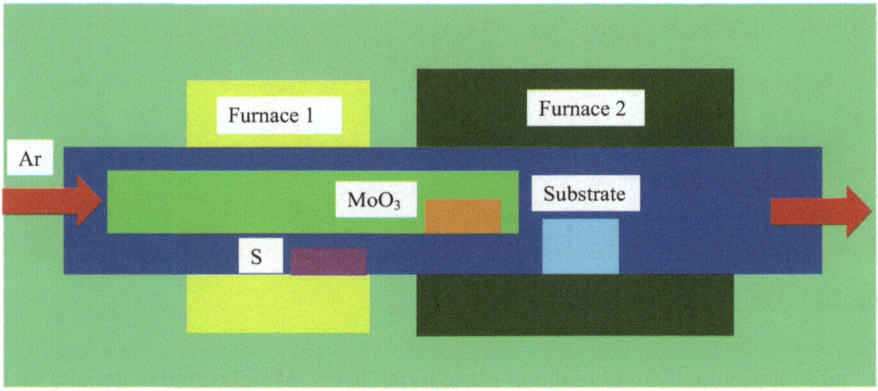

Fig. 2.24 vdW growth of the MoS_2/h-BN heterostructure

furnace 2, and the temperature is tuned by modifying the distance between MoO_3 and the center of the furnace. The h-BN/SiO_2 chip is placed in the center of furnace 2. The growth temperature for S was 180 °C, for MoO_3 was 300 °C and for the substrate 800 °C. The morphology, number of layers, and strain and optical properties were studied with the help of Raman and PL spectra. However, because no electrical measurements were performed, it is not clear if h-BN has the same influence as in the case of graphene, i.e. increases the mobility and the mean-free-path.

The method of vdW epitaxy is used also for the growth of MoS_2/graphene heterostructures (Miwa et al. 2015). In this case, graphene is grown first on the SiC substrate, and the subsequent growth of MoS_2/graphene heterostructure is achieved via PVD. More precisely, Mo is deposited starting from a Mo solid source mounted in an electron beam evaporator, and the substrate is heated at 1050 K in a low-pressure H_2S atmosphere. The entire growth process takes place in situ in a UHV chamber. A cycled method is used to get a coverage of 0.55 ML, which consists of mainly monolayer MoS_2 islands on graphene with Mo clusters on top of them, while at 0.85 ML coverage, about 40% of the islands are bilayer MoS_2.

WS_2/MoS_2 vertical and in-plane heterostructures can be grown in a single step via PVD using a precise control of temperature (Gong et al. 2014). In this case, as shown in Fig. 2.25, MoO_3 powder is located in front of the SiO_2/Si chip, on the surface of which W and Te are spread over; Te is added because it accelerates the melting of W. S powder is placed at a larger distance from the SiO_2/Si chip, in a zone with lower temperature. The carrier Ar gas transports S vapors towards the chip, on which MoS_2 and WS_2 grow sequentially and not as a mixture/alloy due to their different nucleation and growth rates. When vertical heterostructures are required, the growth is performed at 850 °C, while lateral heterostructures are grown at 650 °C. Raman and PL are used to characterize the heterostructures, which have a size of 20–30 μm, and to verify their uniformity.

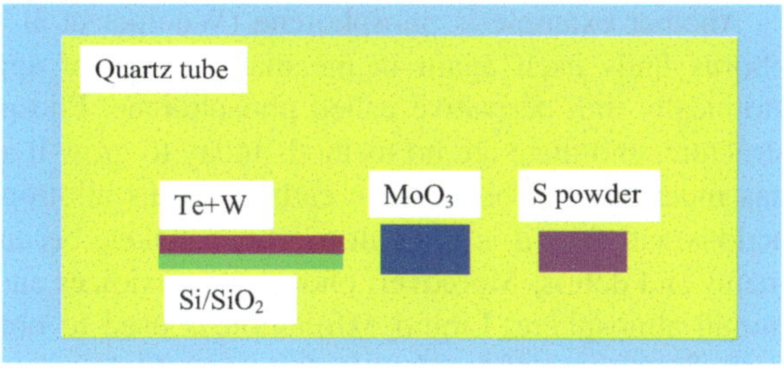

Fig. 2.25 PVD growth of the WS_2/MoS_2 heterostructure

2.1.4 The Growth of X-nes

As mentioned before, **X-nes** are atomically thin materials containing atoms of a single element arranged in a honeycomb lattice, such as graphene (carbon atoms), silicene (silicon atoms), germanene (germanium atoms), and phosphorene (phosphorus atoms).

Silicene, for instance, is a monolayer of a buckled honeycomb lattice of sp^2/sp^3 hybridized Si atoms, which has similar properties as graphene (Vogt et al. 2012). Silicene does not exist in nature, and therefore there are no atom-thick flakes available, as in the case of graphene and TMDs. As such, silicene is grown by depositing Si in high vacuum conditions at 220–260 °C on very clean Ag(111) surfaces prepared by Ar^+ ion bombardment, followed by annealing at 530 °C. The structural and physical properties of silicene were demonstrated by techniques such as ARPES and STM. It is interesting that when Si is epitaxially grown on Ag(110), a multitude of parallel silicon nanoribbons form (Le Lay et al. 2009), all having the width of 1.6 nm (see Fig. 2.26). They are in fact arch-shaped silicene stripes. Similar silicene nanoribbons are obtained if Si is grown on Ag(100).

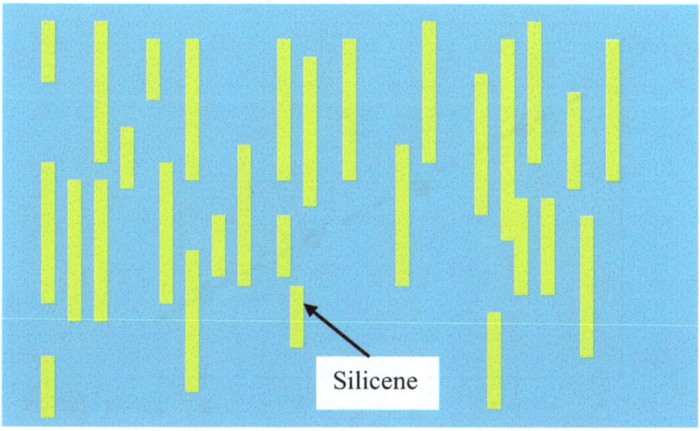

Fig. 2.26 Silicene nanoribbons on Ag(110)

Another example is phosphorene (Woomer et al. 2015). The bulk black phosphorus finds itself again in the main stream of applied research due to its 2D atomically thin derivative called phosphorene. Phosphorene displays a corrugated structure, and there are no methods today to grow it at the wafer scale. Phosphorus has much weaker bonds than carbon, forms allotropes, shows quite strong interactions and thus it is difficult to obtain flakes, because the outcome is only fragments and debris. Moreover, phosphorus oxidizes and becomes phosphoric acid in humid atmosphere. Liquid exfoliation is used to obtain phosphorene monolayers with dimensions of few hundreds of nm. Flakes dispersed in various solvents are produced at 10-g-scale using the shear mixing method (Woomer et al. 2015). Plasma-assisted fabrication of phosphorene by mechanical cleavage followed by Ar^+ thinning is another method to obtain this material. Raman analysis is used to determine the thickness of phosphorene (Lu et al. 2014).

Based on these incipient systematic growth methods, it was demonstrated that phosphorene is a semiconductor having a bandgap that varies with the number of layers, as displayed in Fig. 2.27. However, there are discrepancies between theory and bandgap values determined by various experimental methods (Woomer et al. 2015). Figure 2.27 suggests the possibility of bandgap-engineering by stacking black phosphorus with different atomic thicknesses, but presently the controlled growth of 2D materials is a difficult task. By far, graphene growth is the most mature technology compared to any other 2D material growth method. Moreover, CVD growth of graphene monolayers at 4- and 6-in. wafer levels is dominating graphene growth procedures (Zhang et al. 2013), although epitaxial growth of graphene on SiC is also a successful method. The CVD growth of graphene is commonly performed on Ni and Cu.

CVD of graphene on Ni is based on annealing at 900–1000 °C, followed by exposure to a mixture of gases containing H_2 and CH_4, when carbon dissolves in Ni to form a solid Ni-C solution, which is cooled in Ar gas. During cooling, the C

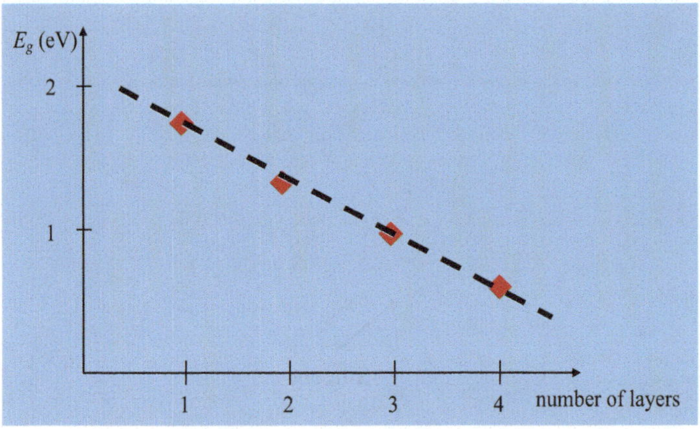

Fig. 2.27 Phosphorene bandgap dependence on the number of layers (after Woomer et al. 2015)

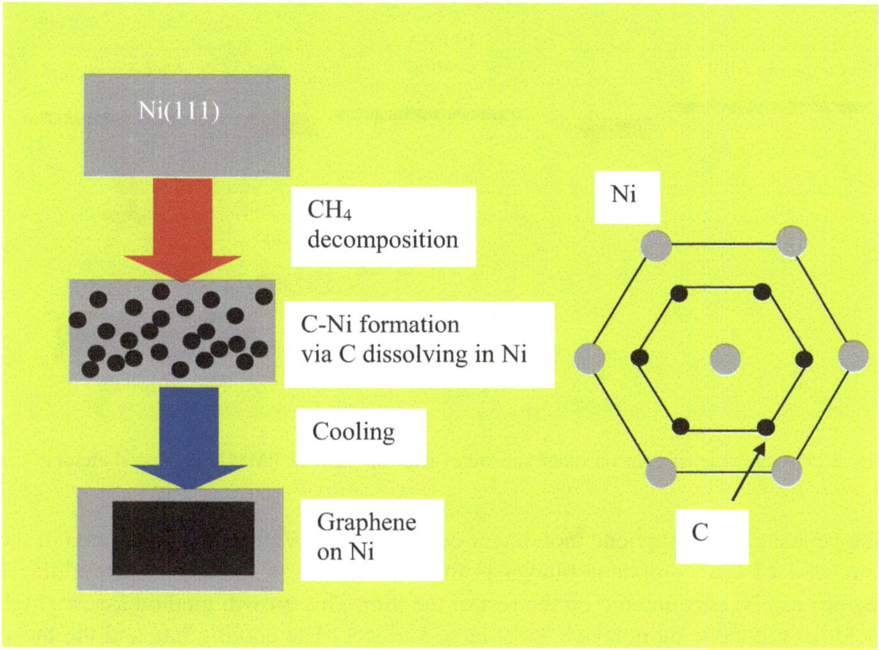

Fig. 2.28 Graphene CVD on Ni(111) and comparison between graphene and Ni(111) lattices

atoms diffuse from the Ni-C solution and form graphene films on Ni, as suggested by Fig. 2.28. Ni(111) is a preferred substrate because it has a hexagonal lattice, similar to graphene.

There are many factors that must be carefully taken into consideration to obtain graphene monolayers of high quality, such as the cooling rate and the Ni film structure. Multilayer graphene regions form due to defects in the Ni microstructure, annealing at high temperatures being able, nevertheless, to alleviate this problem. After the growth process is finished, graphene is transferred to a desired substrate, a technological step that can introduce additional defects in the graphene structure. Ni is a good substrate, but the growth of graphene and its uniformity is limited critically by the grain size of Ni after annealing. Different recipes for graphene growth on Ni exist (see Table 1 in Zhang et al. (2013)), and large areas of graphene on Ni are commercially available today.

Various metals were tried in order to overcome the problems encountered in CVD graphene growth on Ni. Among them, Cu is presently the metal used to grow graphene in batch fabrication at 4-in. and even 6-in. wafer scale. Initially (Li et al. 2009), graphene was grown on a thin Cu foil, with a thickness of 25 μm. Cu foils are annealed at 1000 °C and, similar to graphene growth on Ni, a gas mixture composed of H_2 and CH_4 is introduced in the hot wall furnace. When a continuous graphene layer is formed, the film is cooled at room temperature. Graphene wrinkles occur at Cu steps as a result of the difference between thermal expansion of

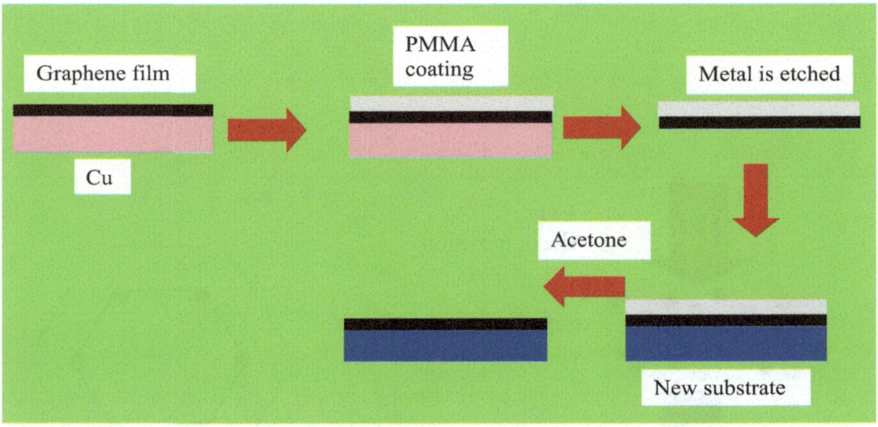

Fig. 2.29 Graphene transfer on other substrates with the help of PMMA and metal etching

graphene and Cu. Graphene monolayer occupies about 95% of the entire area of the film, of 1×1 cm^2, graphene bilayer is found on 3–4% of the area, and multilayer regions can be encountered on the rest of the film. This growth method assures high yield for graphene monolayers, over large surfaces. The cooling rate and the metal thickness play only a minor role in CVD graphene growth on Cu. The segregation process that occurs in the case of graphene growth on Ni is not present when graphene is grown on Cu, since Cu has a very low rate of carbon solubility. Cu is a good catalyst, which decomposes the hydrocarbon and, after the graphene monolayer is deposited over the entire area of the Cu foil, the process is stopped since there is no longer any catalyst. Again, various growth recipes are found in (Zhang et al. 2013).

The transfer of graphene on various substrates can be made in many ways, but a straightforward method is indicated in Fig. 2.29. The transfer method is applicable at the wafer level for solid substrates such as Si/SiO_2 and flexible substrates such as PET.

2.2 Physical Properties of Two-Dimensional Materials

Because hundreds of 2D materials have been discovered in the last years, their physical properties are not fully investigated. There are many theoretical papers, which compute these properties, but the experimental work is at the beginning. In this section, we focus mainly on MoS_2 and some other TMDs, because their mechanical properties are most studied.

The semiconducting MoS_2 monolayer has a thickness of 0.65 nm and is formed from a plane of Mo atoms arranged in a honeycomb lattice, placed between two S planes with S atoms arranged also in a hexagonal lattice (Ganatra and Zhang 2014).

The S-Mo-S covalent bond has a length of 0.24 nm, while the lattice constant is 0.32 nm. The hexagonal structure of the semiconducting MoS_2, analogous to that of graphene, is replaced by an octahedral structure in the case of metallic MoS_2.

2.2.1 *Traps and Defects*

Simple electrical measurements are used to characterize the electronic properties of 2D materials such as MoS_2 (Zhu et al. 2014). In particular, current-voltage and capacitance-frequency (*C-f*) dependences are able to provide information about traps with different time constants in top-gate FETs with MoS_2 monolayer as channel. The *C-f* characteristics measured at various gate voltages in a device with a channel width of 4 μm and a length of 44 μm indicate the presence of different types of traps, with different time constants, via humps in *C-f* dependences. This is a typical behavior not only for MoS_2, but for any semiconductor, and the *C-f* method is extensively used for semiconductor characterization. In the case of the FET analyzed in (Zhu et al. 2014), a double hump was observed at a certain top gate voltage denoted by V_{TG}, as shown in Fig. 2.30.

By denoting the two traps as T1 and T2, the impedance of a circuit model containing the quantum capacitance C_q in parallel with the parasitic capacitance C_p, both in series with the gate oxide capacitance C_{ox}, and further in parallel with the admittances of the two defects, labeled Y_{T1} and Y_{T2}, we have

$$Z(\omega) = (Y_{T1} + Y_{T2} + i\omega C_q + i\omega C_p)^{-1} + (i\omega C_{ox})^{-1} + r_s. \tag{2.2}$$

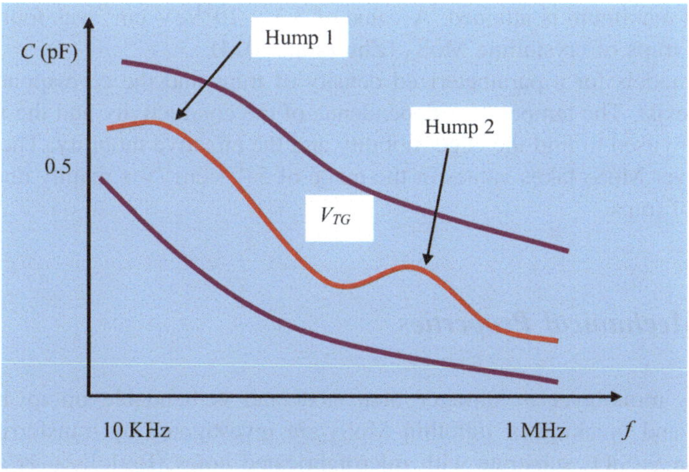

Fig. 2.30 Evidence of traps in *C-f* measurements

In (2.2), r_s is the series resistance of the transistor and

$$Y_{Ti} = (R_{Ti})^{-1} + i\omega C_{Ti} = [\tau_{Ti}/C_{Ti} + 1/(i\omega C_{Ti})]^{-1}, \quad (2.3)$$

where C_{Ti} and τ_i are the capacitances and the time constants of the traps Ti, $i = 1, 2$, and R_{Ti} are the corresponding resistances. The trap capacitance is a direct measure of the trap density via $C_{Ti} = eD_{Ti}$.

Considering that what is really measured is a capacitance C in series with a resistance R, the measured capacitance is related to $Z(\omega)$ as

$$C = -(\omega \mathrm{Im}[Z])^{-1}. \quad (2.4)$$

There is good agreement between the model and measurements (Zhu et al. 2014).

An alternative way to measure the trap densities and their time constants uses the AC conductance obtained from measurements of the capacitance and resistance in frequency at various gate voltages (Zhu et al. 2014). The conductance is written as

$$G(\omega)/\omega = eD_{Ti} \ln[1 + (\omega\tau_{Ti})^2]/(2\omega\tau_{Ti}) \quad (2.5)$$

from which the density of traps and their time constants can be extracted as

$$D_{Ti} = (2.5/e)(G(\omega)/\omega)_{\max} \quad (2.6)$$

and

$$\tau_{Ti} = 1.98/2\pi f_0 \quad (2.7)$$

where $(G(\omega)/\omega)_{\max}$ is the maximum value of $G(\omega)/\omega$ and f_0 is the frequency at which this maximum is attained. A value of 3.3×10^{14}/eV cm^2 was found for the density of traps of crystalline MoS_2 (Zhu et al. 2014).

Other models for a parameterized density of traps and the corresponding time constants exist. The temperature dependence of the conductivity and the activation energies are used to find the edge mobility and the effective mobility. The mobility in monolayer MoS_2 takes values in the range of 5–25 cm^2/V s mainly due to large densities of traps.

2.2.2 Mechanical Properties

The MoS_2 monolayer is stronger than steel and is bendable up to 11%. The stretching and breaking of ultrathin MoS_2 are investigated by transferring MoS_2 flakes on a Si/SiO_2 substrate with microfabricated holes (Bertolazzi et al. 2011). A 270 nm thickness of SiO_2 is optimum for a maximum visibility of MoS_2

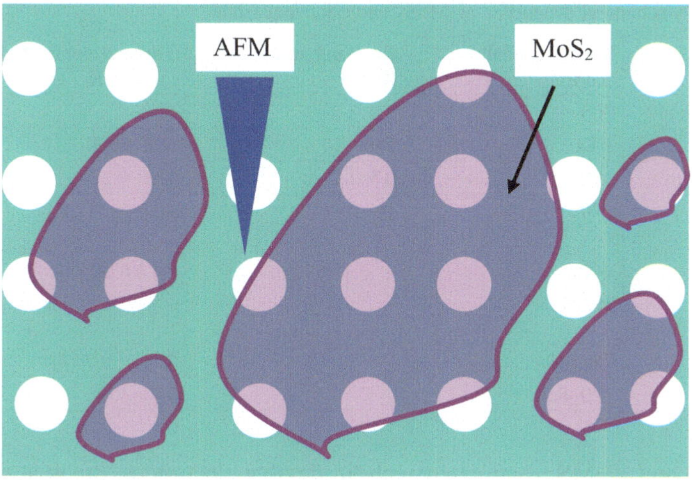

Fig. 2.31 MoS_2 mechanical properties testing

(Benameur et al. 2011). These suspended membranes are then actuated by the tip of an AFM for deformation and breaking, as shown in Fig. 2.31.

The cantilever deflection is monitored by AFM when the probe is moving vertically with a speed of 2 μm/s, deflecting the MoS_2 membrane with mechanical forces up to 150 nN. MoS_2 is considered a linear elastic material, characterized by the elastic modulus and Poisson ratio.

In the case of a MoS_2 membrane suspended over holes with a diameter of 550 ± 10 nm and deflected by an AFM with a tip radius $r = 12 \pm 2$ nm, the force-deflection dependence is obtained from AFM data fitted with the expression

$$F = \sigma_i \pi \delta + E q^3 \delta^3 / r^2. \tag{2.8}$$

In (2.8), δ is the deflection, σ_i is the prestress of the membrane, E is the Young modulus, and q is related to the Poisson ratio v as

$$q = 1/(1.05 - 0.15v - 0.16v^2) \tag{2.9}$$

where, for MoS_2, $v = 0.27$ and $q = 0.95$. The Young modulus, breaking strength and thermal conductivity of some materials are given in Table 2.3 (Bertolazzi et al. 2011; Akinwande et al. 2014).

Mechanical properties better than those of MoS_2 are encountered only in the case of graphene, which is the strongest known material. MoS_2 has impressive mechanical properties and is by far among the strongest semiconductors known to date.

Considering a linear relation between stress σ and strain ε of the form $\sigma = \varepsilon E$, it results that the breaking point is attained for $\varepsilon_b = 0.06$–0.11, i.e. at 6–11%.

Table 2.3 Mechanical properties of MoS_2 and other materials

Material	Young modulus (GPa)	Breaking strength (GPa)	Thermal conductivity ($Wm^{-1}\,K^{-1}$)
Stainless steel	205	0.9	16
Graphene	1000	130	600–5000
Molybdenum	329	0.5–1.2	138
MoS_2 monolayer	270	16–30	40
MoS_2 bulk	240	–	4 (⊥), 50 (‖)
WSe_2 monolayer	195	–	–
WSe_2 bulk	75–100	–	2 (⊥), 9.7 (‖)
h-BN	220–880	–	250–360

Polyimide and PDMS, used extensively in flexible electronics, break at a strain of 7%, so that the route towards flexible electronics is open for TMDs (Akinwande et al. 2014).

The impressive mechanical properties of MoS_2 monolayers can be used for high-quality and high-frequency NEMS mechanical resonators (Lee et al. 2013). Such a resonator is fabricated from atomically thin MoS_2 directly exfoliated on etched Si/SiO_2, as indicated in Fig. 2.32. Nanomechanical resonators consist of MoS_2 diaphragms with a diameter of about $d = 6$ μm and a thickness in the range 13–68 nm, corresponding to a number of monolayers between 20 and 97. Experiments on a MoS_2 diaphragm of 5.7 μm in diameter and a thickness of 68.1 nm show that the fundamental mode is attained at 19.68 MHz with a quality factor of $Q = 710$ in moderate vacuum. Using an optical interference technique, the displacement sensitivity was found to be 49.5 fm/Hz$^{1/2}$. In the case of a thinner

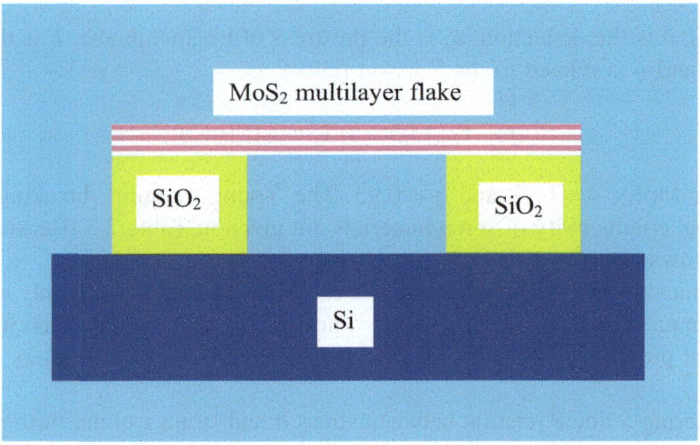

Fig. 2.32 The nanomechanical MoS_2 resonator

MoS_2 diaphragm, with a diameter of 5.5 μm and a thickness t = 38 nm, which corresponds to 54 layers, the resonance frequency attains 14.13 MHz, Q = 550 and the displacement sensitivity is 33 fm/Hz$^{1/2}$. The thinnest and smallest diaphragm is 9-layer thick and has d = 1.9 μm, a resonance at 49.7 MHz and a quality factor of 80. The quality factor of these resonators scale with the thickness, while the surface (S) to volum (V) ratio, $S/V \approx 1/t$. Moreover, the figure of merit FOM of the mechanical resonator, expressed by

$$FOM = f_0 \times Q, \tag{2.10}$$

reaches the highest value 2 × 10^{10} Hz, which overcomes that of graphene NEMS resonators working in similar conditions.

The quality factors drop dramatically, up to 0.1, when the resonators are placed in air. The air damping, i.e. the dependence of Q on the pressure p, can be expressed as

$$Q \propto p^{-1/2} \quad \text{in the pressure range } 1-100\,\text{Torr} \tag{2.11a}$$

$$Q \propto p^{-1} \quad \text{in the pressure range } 100-1000\,\text{Torr} \tag{2.11b}$$

The first tunable NEMS resonators consisting of mono- and few-layer MoS_2 were reported recently in (Feng et al. 2014a) and were tested with the circuit illustrated in Fig. 2.33. In particular, the nanomechanical resonator with a four-monolayer MoS_2 diaphragm has two resonant frequencies located in the radio domain, at 77 and 106 MHz, with a signal level of almost 100 μV. The frequency

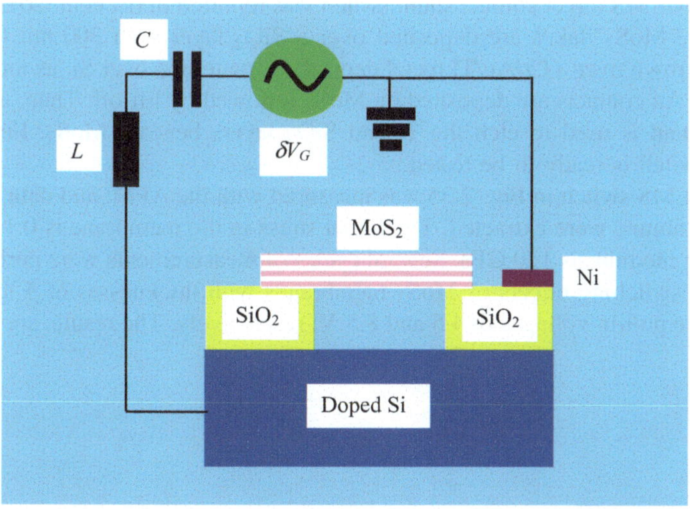

Fig. 2.33 Tunable NEMS MoS_2 mechanical resonator

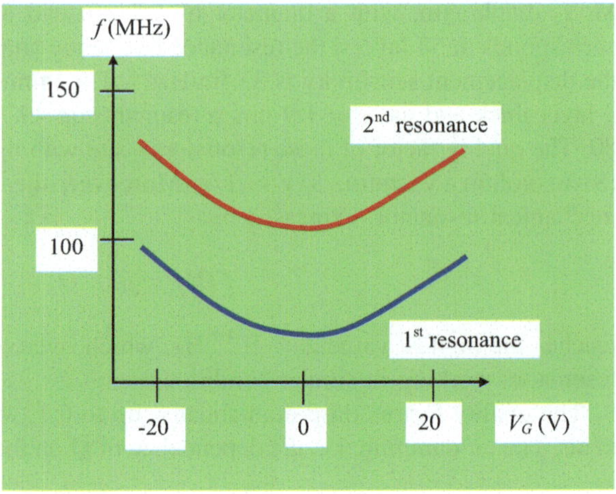

Fig. 2.34 Tuning the resonance frequency of the nanomechanical MoS_2 resonator via backgate

of the NEMS resonator can be tuned with an applied DC backgate voltage, as shown in Fig. 2.34. This figure suggests a very large tunability, of

$$\Delta f/f = 84\,\text{ppm}/\text{single electric charge}. \tag{2.12}$$

The dynamic range is also a *FOM* for NEMS, indicating high sensitivity and low phase noise. The dynamic range is defined as the ratio in dB between the value of the signal at which the nonlinear response regime is reached and the thermo mechanical noise (Brownian motion). This parameter is in the range 70–110 dB for the NEMS resonators in (Feng et al. 2014a), which is higher than in other 2D mechanical resonators.

A first MEMS direct contact shunt switch was reported in (Li et al. 2013). In this device, the MoS_2 flakes are deposited over a SiO_2 layer with 300 nm thickness, which is grown over a Cr/Au/Ti metal deposition, patterned over Si, as indicated in Fig. 2.35. Au contacts are deposited on MoS_2 followed by lift-off. Then, a buffered oxide etchant is used to etch the Ti and SiO_2 layers beneath MoS_2. Finally, the MEMS switch is ready to be tested.

The MEMS switch in Fig. 2.35 was measured with the AFM, and data about the MoS_2 membrane were extracted. The initial stress in the membrane is 0.1 GPa and the Young modulus is 170 GPa. Several cycles of measurements were performed to verify the switching effects of MoS_2 membranes with thicknesses of 5 and 8 nm, which have pull-in voltages of 4.5 and 8.5 V, respectively. The results are shown in Fig. 2.36.

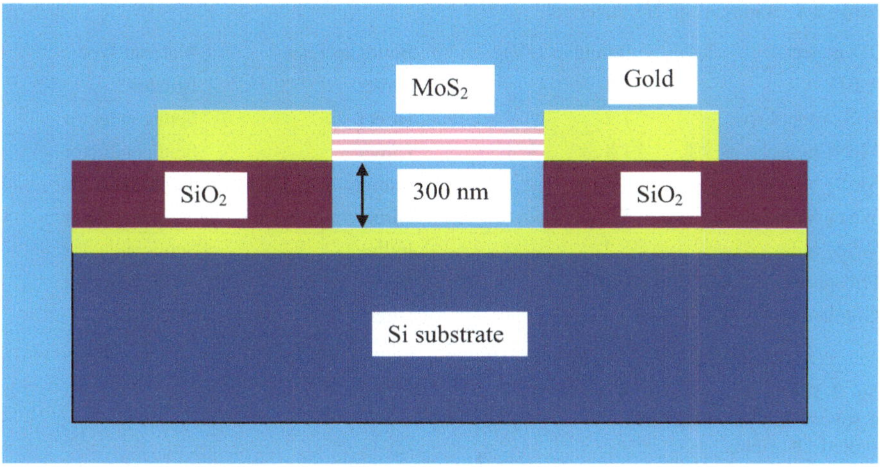

Fig. 2.35 MEMS switched based an atomically thin MoS_2 membrane

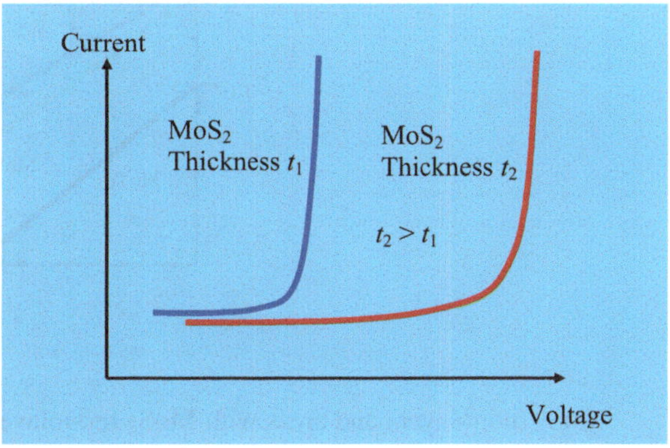

Fig. 2.36 MEMS shunt switch based on MoS_2 membranes actuation

2.2.3 Electronic Properties

The bandgap values and types of the main 2D materials are presented in Table 2.4. (Akinwande et al. 2014; Kanazawa et al. 2016). The bandgap of 2D materials can be tuned by various parameters, as described below.

(a) *Bandgap tunability due to applied DC voltages*
The bandgap of bilayer TMDs can be tuned via applied DC fields (Ramasubramaniam et al. 2011), as indicated in Fig. 2.37. The same effect is encountered in graphene bilayers.

(b) *Bandgap tunability due to different layer numbers*
Different atomically thin semiconductors have different bandgaps, depending on the number of monolayers involved. In recent experiments (Huang et al. 2015),

Table 2.4 Bandgap of 2D materials

2D material	Bandgap (eV)	Bandgap type	Material type
h-BN	5.9	Direct	Isolator
HfS_2 monolayer	1.2	Direct	Semiconductor
MoS_2 monolayer	1.8	Direct	Semiconductor
MoS_2 bulk	1.2	Indirect	Semiconductor
WSe_2 monolayer	1.7	Direct	Semiconductor
WSe_2 bulk	1.2	Indirect	Semiconductor
Phosphorene	0.5–2	Direct	Semiconductor
Graphene	0	Direct	Semimetal

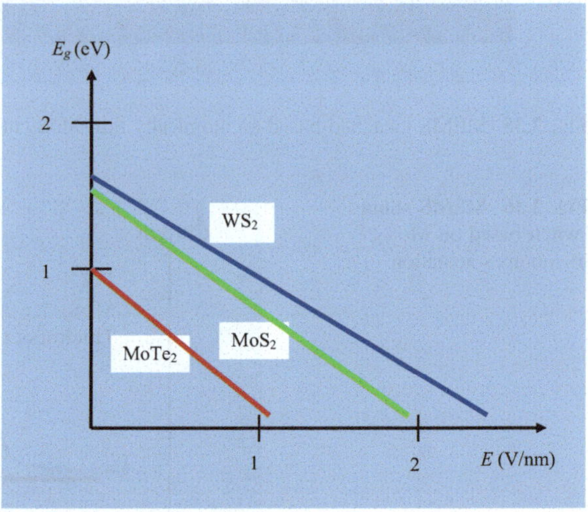

Fig. 2.37 Bandgap variation of several bilayer TMDs via applied DC fields

MoS_2 monolayers and areas with MoS_2 monolayers (ML), bilayers (BL) and trilayers (TL) were grown by CVD. STM analysis was used to study each type of MoS_2 by measuring the height profile of the material under study. The bandgap was determined from dI/dV data as a function of the bias voltage, its value being determined by the voltage interval where dI/dV vanishes, as shown in Fig. 2.38. The bandgap of MoS_2 as a function of the thickness/layer number is summarized in Table 2.5. Note that grain boundaries are also a source of bandgap tuning. Close to the grain boundary, the bandgap in ML MoS_2 can decrease with as much as 0.85 eV, recovering its value in Table 2.5 at distances higher than 4 nm from the boundary.

(c) *Bandgap tunability due to strain*

There are various ways to strain MoS_2. For example, wrinkles appear on MoS_2 monolayers when transferred on an elastomer, which is stressed 100% and then released. The Raman and PL spectra are thus shifted, the shift indicating the strain-induced tuning/decrease of the bandgap (Castellanos-Gomez et al. 2013).

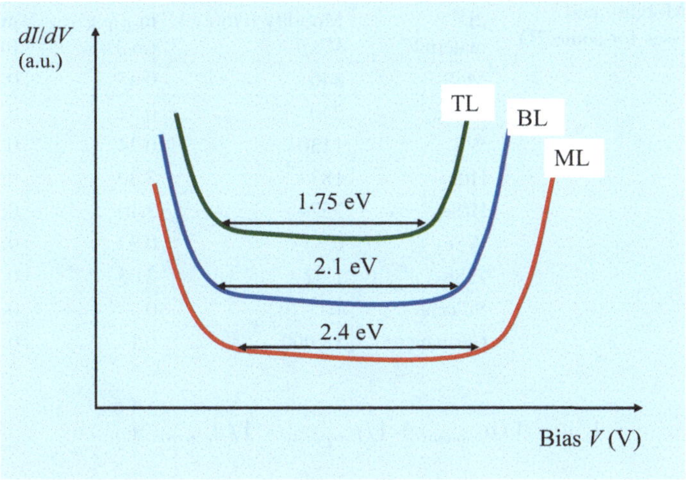

Fig. 2.38 Bandgap tuning of MoS_2 as a function of thickness. The curves are offset for clarity

Table 2.5 MoS_2 bandgap as a function of its thickness

MoS_2 number of layers	Bandgap (eV)
ML	2.40
BL	2.10
TL	1.75

The best method to tune the bandgap depends ultimately on the 2D material under study. For example, the strain can induce a bandgap in graphene, but its value is rather small. On the other hand, the DC fields required to tune the bandgap at desirable values could be too high and could harm the tiny monolayers. The only viable way to tune this parameter is to obtain van der Waals homostructures, formed from regions with an appropriately chosen number of monolayers (for example monolayer/trilayer or bilayer/trilayer), or heterostructures formed from different 2D materials. In many applications, such as sensing, bandgap tuning need not be large. For instance, in many strain detectors this parameter is typically tuned with about 100 meV by a strain of 2.5%.

Mobility and effective masses are also key parameters for 2D materials. We have already presented some experimentally extracted mobility values in the section dedicated to the growth of 2D materials. For many 2D materials the mobility and effective masses are not even calculated yet. Recently, these parameters were computed for some important atomically thin semiconductors using full-potential local-orbital codes based on the local density approximation (LDA). In particular, the calculation of mobility must take into account various scattering mechanisms: on phonons, impurities, electrons, and so on, the total mobility being described by the relation (Zhang et al. 2014).

Table 2.6 Mobility and effective masses for some 2D materials

2D material	Mobility (cm^2/V s)	$m_{eff\ \Gamma\text{-}K}$ (m_0)	$m_{eff\ K\text{-}M}$ (m_0)
MoS_2	340	0.45	0.45
$MoSe_2$	240	0.52	0.52
WS_2	1130	0.24	0.26
HfS_2	1833	3.30	0.24
$HfSe_2$	3579	3.10	0.18
WSe_2	0.33	0.33	0.31
$ZrSe_2$	1.62	2.03	0.22
Silicene	10^3–10^5	0	0
Germanene	18,000	0.1	0.1

$$1/\mu = 1/\mu_{phonons} + 1/\mu_{impurity} + 1/\mu_{electron} + \ldots \tag{2.13}$$

Because the most important scattering mechanism for electrons is that on phonons, the mobility is phonon limited. More precisely, it is limited by long wave acoustic phonons. The LDA method was used to determine the structural, electronic and mechanical parameters of 2D materials. Some values of these parameters are presented in Table 2.6 (Zhang et al. 2014; Schwierz et al. 2015).

As can be seen from this table, there are 2D materials such as $HfSe_2$ and HfS_2 with impressive mobilities and electron masses along the Γ-M direction. For transistor channels with a drain-source distance of 5 nm, the tunneling current becomes high for high mobility semiconductors with low effective masses, so that 2D materials with high mobilities and high effective masses could solve in future the short channel effects' conundrum. The experiments are, however, in infancy since many values in Table 2.6 are only calculated and not measured yet.

Similarly, band offsets of various 2D materials, especially TMDs and heterostructures based on them, are studied mostly with ab initio simulation packages (Kang et al. 2013; Gong et al. 2013). Such simulations determine many physical properties, the most important being the bandgap, the conduction and valence bands edges, and the workfunction of each TMD. Examples are provided in Table 2.7 (Gong et al. 2013).

Of course, depending on the software, there are differences between various numerically simulated parameters and the corresponding experimental values.

Table 2.7 Conduction and valence bands edges, bandgaps and work functions of TMDs

2D material	E_v (eV)	E_c (eV)	E_g (eV)	Type of gap	Work function (eV)
HfS_2	−6.33	−5.1	1.23	Indirect	−5.71
$HfSe_2$	−5.39	−4.94	0.45	Indirect	−5.17
WS_2	−5.5	−3.96	1.54	Direct	−4.73
WSe_2	−4.87	−3.54	1.32	Direct	−4.21
MoS_2	−5.86	−4.27	1.59	Direct	−5.07
$MoSe_2$	−5.23	−3.9	1.32	Direct	−4.57

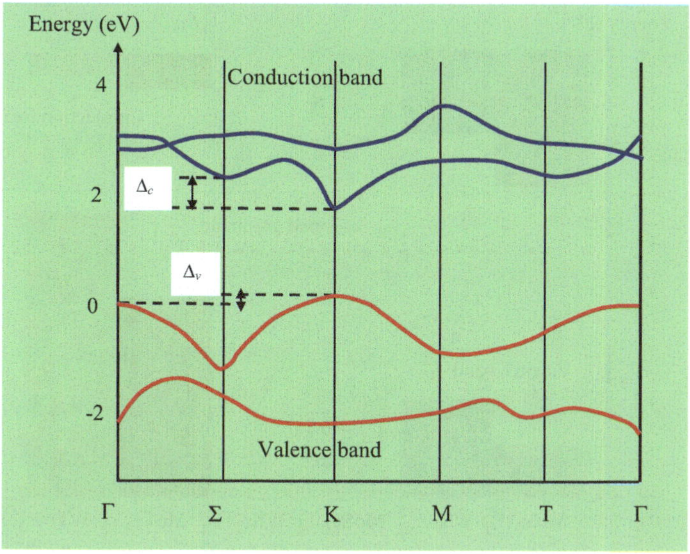

Fig. 2.39 Significance of Δ_c and Δ_v in a TMD with the same bandstructure as MoS_2

Table 2.8 Conduction and valence band off-sets calculated by DFT and GW methods

TMD	Δ_c (eV)	Δ_v (eV)
MoS_2	0.25; 0.23	0.02; 0.16
$MoSe_2$	0.23; 0.33	0.23; 0.34
WS_2	0.25; 0.25	<0.002; 0.06
WSe_2	0.21; 0.36	0.26; 0.34

First-principles and several numerical approximations are also used to compute the band edges and band offsets of monolayers of Mo- and W-based dichalcogenides (Liang et al. 2013). Some values of energy differences between minima and maxima in the conduction and valence bands at different points in the first Brillouin zone, denoted Δ_c and Δ_v, respectively, in Fig. 2.39, are presented in Table 2.8, where the first value was calculated by using the density functional theory (DFT) and the second one by the GW method (Liang et al. 2013).

The band energy diagram for several TMDs is displayed in Fig. 2.40; the values of conduction and valence band edges differ in several papers depending on the method of calculation/measurement. However, this figure suggests that we can engineer various homo/heterostructures, all semiconductors in Fig. 2.40 having a direct bandgap. In the case of heterostructures, we can choose two different 2D semiconductors from those in Fig. 2.40 or other 2D materials that can be found in (Kang et al. 2013; Gong et al. 2013; Liang et al. 2013). In general, these heterostructures are vertical. In the case of homostructures, we rely on the fact that the bandgap of 2D semiconductors varies with their thickness, as discussed above. So, homojunctions, such as monolayers/bilayers, trilayers/monolayers, etc., are

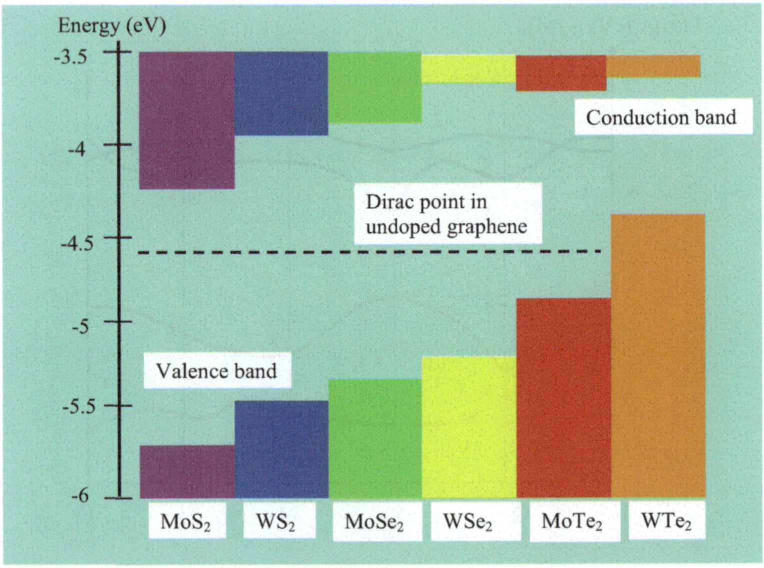

Fig. 2.40 The band diagram of several TMDs

possible in configurations with lateral geometries, called lateral heterostructures. These considerations can be applied, for example, to homostructures in phosphorene, where the bandgap varies strongly with the number of layers (Cai et al. 2014). Using fitting parameters, it can be shown that the bandgap of phosphorene containing few atomic layers is given by

$$E_g = E_0 + C/n^{\gamma} \tag{2.14}$$

where E_0 = 0.39 eV is the bandgap of bulk black phosphorus, n is the number of layers, while γ = 1.4 and C = 1.62 eV. Relation (2.14) does not hold for phosphorene monolayer, due to absence of the interlayer coupling.

Similarly, the calculated maximum value of the workfunction of phosphorene (depending on the simulation method) varies with its thickness, from about 5.2 eV for a monolayer, 4.7 eV for a bilayer, to 4.5 eV for several layers. The band structure of phosphorene is represented in Fig. 2.41.

The band offsets and edge states in lateral TMDs homostructures such as bilayer-monolayer $MoSe_2$ or WSe_2 can be visualized by STM (Zhang et al. 2015). Lateral homostructures are obtained when another layer of the same TMD is grown on the top of the monolayer via CVD or MBE. An example of such a lateral homostructure is represented schematically in Fig. 2.42.

STM reveals the band structure of lateral homostructures as well as the band bending near the interfaces, as displayed in Fig. 2.43. The corresponding parameters of WSe_2 and $MoSe_2$ lateral homostructures are Δ_{VBO} = 0.12 and 0.43 eV,

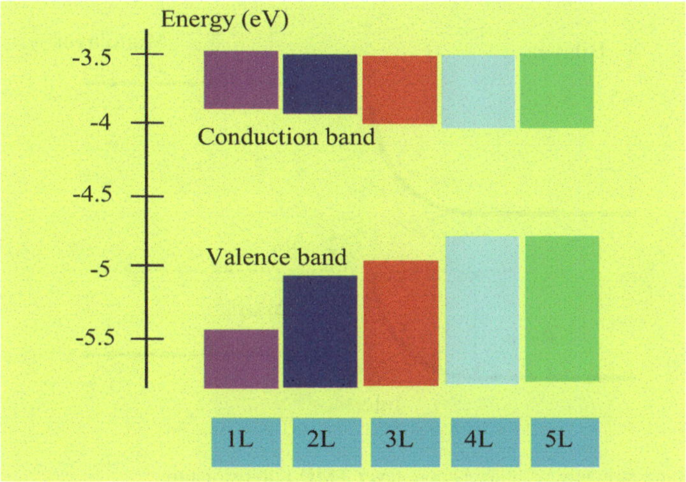

Fig. 2.41 Thickness-dependent energy band structure in phosphorene in lateral heterostructures

Fig. 2.42 Lateral homostructure in TMDs

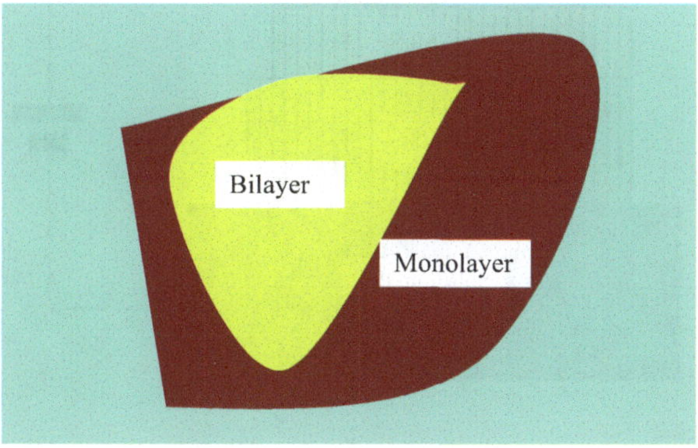

$\Delta_{CBO} = 0.15$ and 0.08 eV, $\Delta_{int} = 0.8$ and 0.4 eV, respectively. In both cases, $\Delta_{bend} = 0.15$ eV.

A vertical heterostructure with vertically aligned layers, as that shown in Fig. 2.44, where the van der Waals planes are perpendicular to the substrate, could be formed from 2D MoS_2 and WSe_2 layers via the kinetically controlled rapid sulfurization and selenization growth method (Yu et al. 2015). MoS_2 is intrinsically an *n*-type semiconductor, while WSe_2 is of *p* type. The entire structure is in fact a vertical *p-n* junction, which shows strong rectification, an on-off ratio of 150, and an ideality factor of 1.5 at voltages <1.5 V. Hopping of charge carriers across polycrystalline domains within MoS_2 and WSe_2 layers is the main transport mechanism of this diode.

Another diode-like vdW heterostructure forms between an *n*-MoS_2 monolayer and a *p*-WSe_2 monolayer (Fang et al. 2014). The monolayers were stacked one over

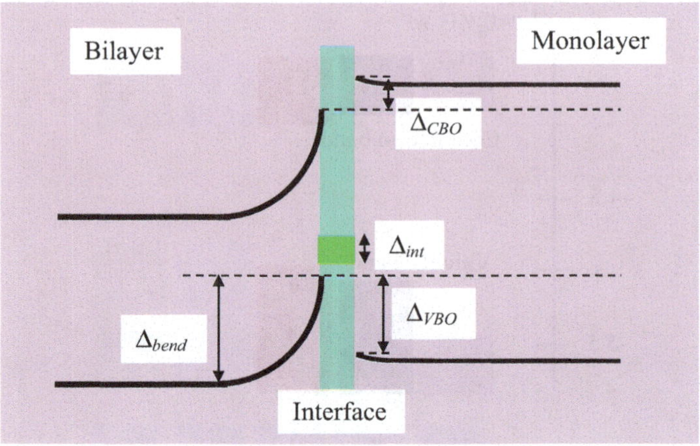

Fig. 2.43 Band structure of bilayer-monolayer TMD homostructure

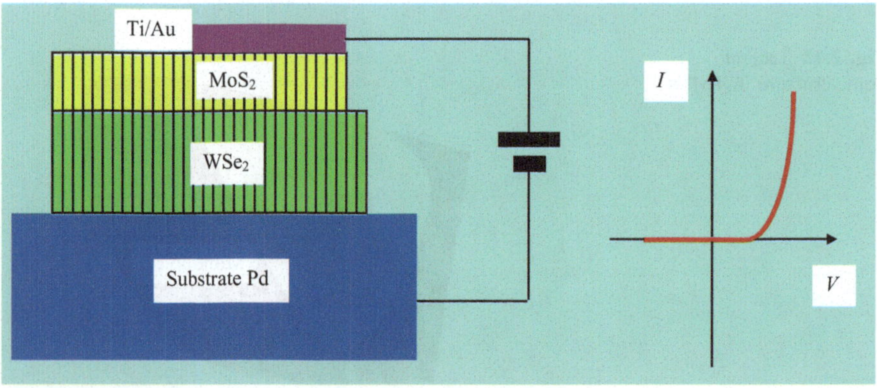

Fig. 2.44 Vertical heterostructure formed from an *n*-type MoS_2 and a *p*-type WSe_2 (*left*) and its $I-V$ characteristic (*right*)

the other on a Si substrate covered by a 260 nm-thick thermally grown SiO_2. Both monolayers display direct bandgaps, and the PL shows two excitonic peaks at 1.64 and 1.87 eV associated to the WSe_2 and MoS_2 monolayers, respectively. In the MoS_2/WSe_2 heterostructure there is a peak at 1.55 eV, at a lower energy than the two peaks corresponding to the individual monolayers, and with 1.5 times higher intensity than that of the MoS_2 monolayer. The band energy of the heterojunction favors the following processes under photo-excitation (see Fig. 2.45): absorption and generation of excitons in individual MoS_2 and WSe_2 monolayers (Abs), relaxation of excitons at the interface of the vdW MoS_2/WSe_2 heterojunction induced by the band offset (Rel), and radiative recombination of excitons (Rec).

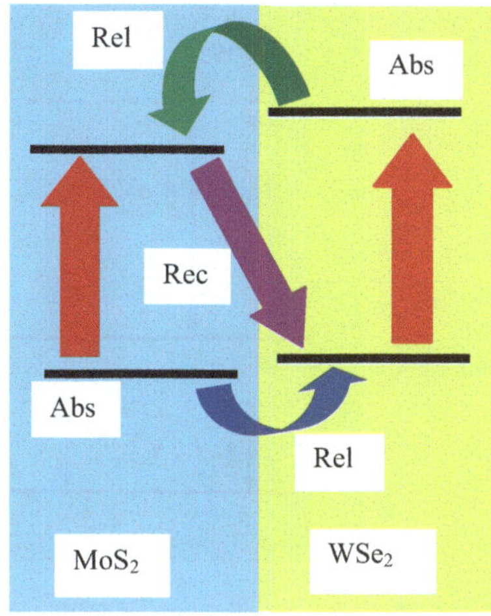

Fig. 2.45 Physical processes in the photo-excited type II heterostructure MoS_2/WSe_2

Other 2D materials and heterostructures based on them can be obtained by MBE growth. An example is the $MoSe_2/HfSe_2$ heterostructure grown on AlN(0001)/Si (111) (Aretouli et al. 2015), which has a type II band alignment. In this case, 6 monolayers of $HfSe_2$ were grown on AlN, and 1, 2, 3, 4, or 8 monolayers of $MoSe_2$ were grown subsequently on top of $HfSe_2$ for structural analysis purposes. The measured work function of $MoSe_2$ is 5.1 eV, while for $HfSe_2$ the value of this parameter is 5.5 eV. The resulting band structure is displayed in Fig. 2.46.

$HfSe_2$ has a very high mobility at room temperature, of 3500 cm^2/V s, and can be grown by MBE (Yue et al. 2015). It forms a sharp interface with the substrate (bulk MoS_2 or highly ordered pyrolytic graphite (HOPG), for example), the flat 2D layers having an octahedral 1T orientation. $HfSe_2$ is an n-type semiconductor with an indirect bandgap of 1.1 eV. Numerous experimental structural analyses were performed to determine the band structure of $HfSe_2$ grown on HOPG, the result being displayed in Fig. 2.47 (Yue et al. 2015).

Van der Waals homojunctions consist, as pointed out above, of multilayered structures of the same material. Because $MoSe_2$ is typically of n type and

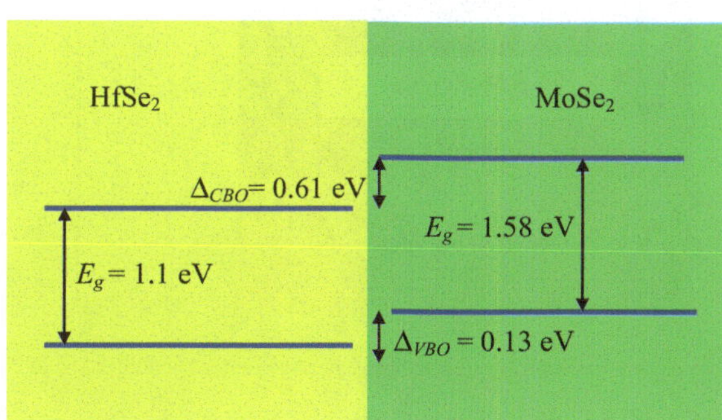

Fig. 2.46 The band structure of the $MoSe_2/HfSe_2$ heterostructure

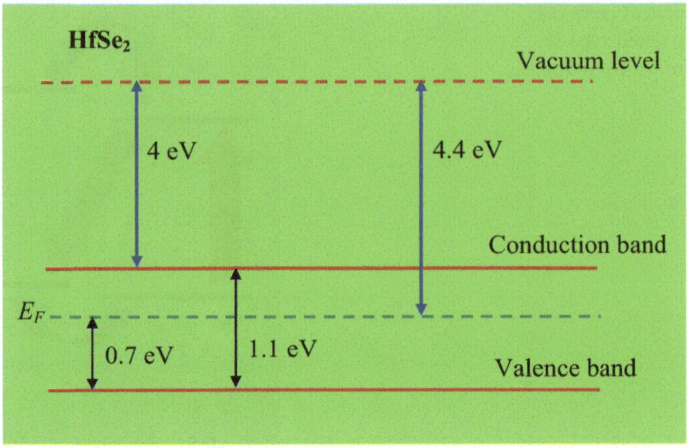

Fig. 2.47 The band structure of $HfSe_2$

transforms in a *p*-type semiconductor by Nb doping, a homostructure acting as tunable *p-n* diode via a backgate voltage can be fabricated from this material. In particular, such a homostructure obtained by overlapping 3.2 nm-thick *n*-MoS_2 on 2.9 nm-thin *p*-MoS_2 already transferred on a SiO_2 substrate was demonstrated in (Jin et al. 2015). Figure 2.48 illustrates schematically such a homostructure as well as its tunable current-voltage characteristic.

As shown in Fig. 2.49, the *I–V* dependence displays four regions: the reverse bias region I, the ideal linear diode region II, the current injection region III, and the series resistance dominant region IV. The *I–V* dependence has the form

$$I = I_S[\exp(V/nV_T) - 1] \tag{2.15}$$

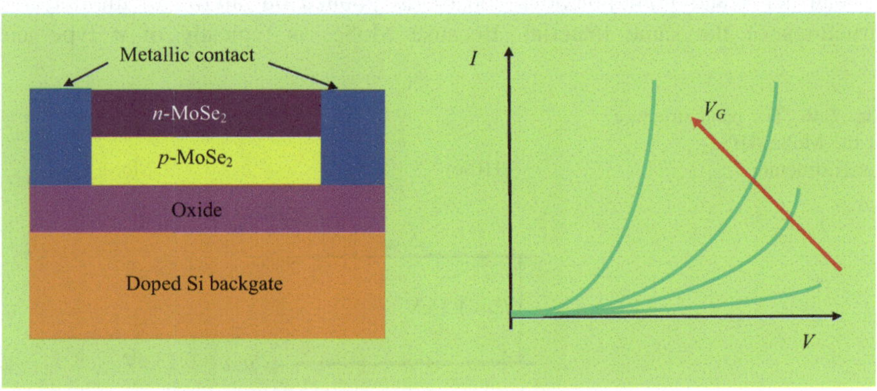

Fig. 2.48 $MoSe_2$ *p-n* diode vdW homojunction and its current-voltage characteristic

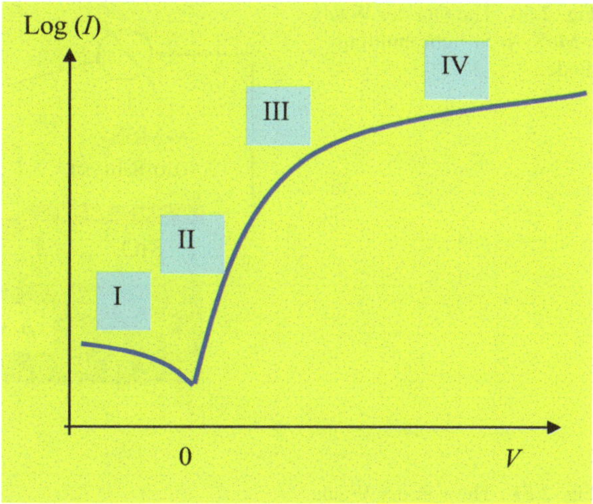

Fig. 2.49 $MoSe_2$ *p-n* diode van der Waals homojunctions

where n is the ideality factor, $V_T = k_B T/e$ is the thermal voltage, and I_S is the saturation current in the reverse bias region. In region II the ideality factor is 1, as in an ideal diode, due to the low density of charge traps, in contrast to bulk semiconductor junctions, where the ideality factor is typically higher than 2. The external quantum efficiency of this vdW homojunction used as photodiode is related to the responsivity R as

$$EQE = Rhc/e\lambda \tag{2.16}$$

and attains a value of more than 1000% at 1 V for a laser power of 890 W/m^2 at the illumination wavelength of λ = 405 nm.

Few-layer $MoSe_2$-based *n*-FET and *p*-FET devices fabricated as above have comparable mobilities, of the order of 10 cm^2/V s. However, the on-off ratio of *p*-FETs, of 10^6–10^7, is with one order of magnitude higher than in *n*-FETs.

Diodes can be formed also from a typical 2D monolayer, such as *n*-type MoS_2, and a 3D semiconductor, such as *p*-type silicon (Lopez-Sanchez et al. 2014). Light emitting diodes and solar cells were produced in this way. A light emitting diode configuration is represented in Fig. 2.50, and resembles similar structures involving graphene and doped semiconductors for solar cells applications. The heterojunction is abrupt, and the band alignment, shown in Fig. 2.51, is of type II staggered (Lopez-Sanchez et al. 2014).

When a forward electric field is applied on the *n*-MoS_2–*p*-Si vdW heterojunction, the electrons from the *n*-MoS_2 monolayer and the holes from *p*-Si recombine radiatively at the interface, and light emission takes place. All the surface area of the heterojunction is radiating with a threshold power density of 3.2 W/cm^2, obtained at a voltage of 5.5 V. The PL spectrum, with a maximum at about 685 nm, contains two peaks associated to excitonic transitions from the bottom of the conduction

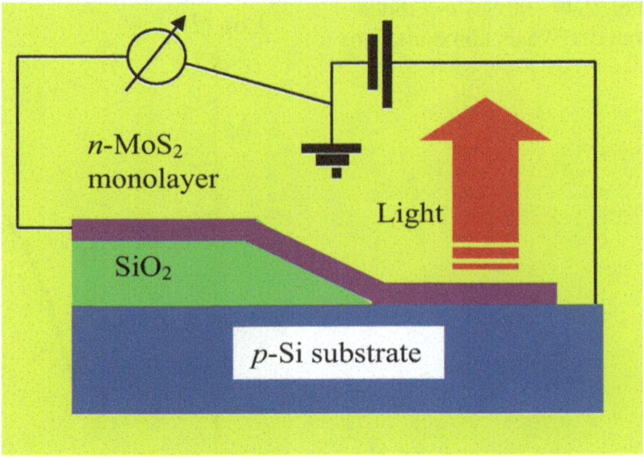

Fig. 2.50 The van der Waals n-MoS_2–p-Si light emitting diode

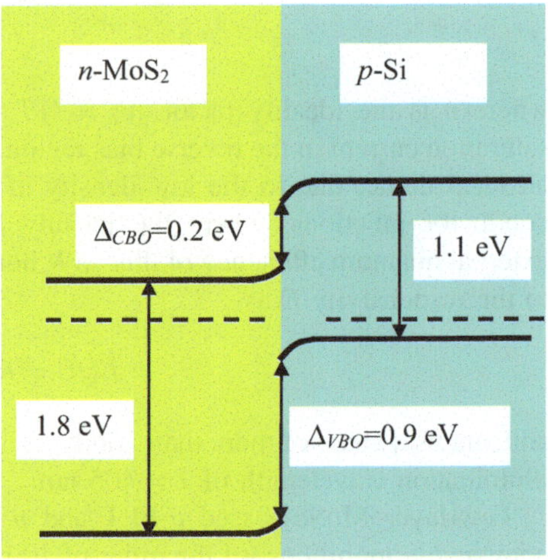

Fig. 2.51 The van der Waals heterojunction n-MoS_2–p-Si

band, E_c, to the top of the valence band, E_v, which are split due to the spin-orbit coupling. The vdW n-MoS_2–p-Si heterojunction works also as a solar cell with an external quantum efficiency of 4.4%.

References

Akinwande D, Petrone N, Hone J (2014) Two-dimensional flexible nanoelectronics. Nat Commun 5:5678

Aretouli KE, Tsipas P, Tsoutsou D, Marquez-Velasco J, Xenogiannopoulou E, Giamini SA, Vassalou E, Kelaidis N, Dimoulas A (2015) Two-dimensional semiconductor $HfSe_2$ and $MoSe_2/HfSe_2$ van der Waals heterostructures by molecular beam epitaxy. Appl Phys Lett 106:143105

Benameur MM, Radisavljevic B, Héron JS, Sahoo S, Berger H, Kis A (2011) Visibility of dichalcogenide nanolayers. Nanotechnology 22:125706

Bertollazi S, Brivio J, Kis A (2011) Stretching and breaking of ultrathin MoS_2. ACS Nano 5:9703–9709

Butler SZ, Hollen SM, Cao L, Cui Y, Gupta JA, Gutiérrez HR, Heinz TF, Hong SS, Huang J, Ismach AF, Johnston-Halperin E, Kuno M, Plashnitsa VV, Robinson RD, Ruoff RS, Salahuddin S, Shan J, Shi L, Spencer MG, Terrones M, Windl W, Goldberger JE (2013) Progress, challenges, and opportunities in two-dimensional materials beyond graphene. ACS Nano 7:2898–2926

Cai Y, Zhang G, Zhang Y-W (2014) Layer-dependent band alignment and work function of few-layer phosphorene. Sci Rep 4:6677

Castellanos-Gomez A, Roldán R, Cappelluti E, Buscema M, Guinea F, van der Zant HSJ, Steele GA (2013) Local strain engineering in atomically thin MoS_2. Nano Lett 13:5361–5366

Chiu K-C, Zhang X-Q, Liu X, Menon VM, Chen Y-F, Wu J-M, Lee Y-H (2015) Synthesis and applications of monolayer semiconductors (June 2015). IEEE J Quant Electron 51:0600100

Chhowalla M, Shin HS, Eda G, Li L-J, Loh KP, Zhang H (2013) The chemistry of two-dimensional layered transition metal dichalcogenide nanosheets. Nat Chem 5:263–275

Coleman JN, Lotya M, O'Neill A, Bergin SD, King PJ, Khan U, Young K, Gaucher A, De S, Smith RJ, Shvets IV, Arora SK, Stanton G, Kim H-Y, Lee K, Kim GT, Duesberg GS, Hallam T, Boland JJ, Wang JJ, Donegan JF, Grunlan JC, Moriarty G, Shmeliov A, Nicholls RJ, Perkins JM, Grieveson EM, Theuwissen K, McComb DW, Nellist PD, Nicolosi V (2011) Two-dimensional nanosheets produced by liquid exfoliation of layered materials. Science 331:568–571

Eda G, Yamaguchi H, Voiry D, Fujita T, Chen M, Chhowalla M (2011) Photoluminescence from chemically exfoliated MoS_2. Nano Lett 11:5111–5116

Eichfeld SM, Hossain L, Lin Y-C, Piasecki AF, Kupp B, Birdwell AG, Burke RA, Lu N, Peng X, Li J, Azcatl A, McDonnell S, Wallace RW, Kim MJ, Mayer TS, Redwing JM, Robinson JA (2015) Highly scalable, atomically thin WSe_2 grown via metal-organic chemical vapor deposition. ACS Nano 9:2080–2087

Fang H, Battaglia C, Carraro C, Nemsak S, Ozdol B, Kang JS, Bechtel HA, Desai SB, Kronast F, Unal AA, Conti G, Conlon C, Palsson GK, Martin MC, Minor AM, Fadley CS, Yablonovitch E, Maboudian R, Javey A (2014) Strong interlayer coupling in van der Waals heterostructures built from single-layer chalcogenides. Proc Nat Am Sci 111:6198–6202

Feng PX-L, Wang Z, Lee J, Yang R, Zheng X, He K, Shan J (2014) Two-dimensional nanoelectromechanical systems (2D NEMS) via atomically-thin semiconducting crystals vibrating at radio frequencies. 2014 IEEE Int Electron Devices Meet (IEDM):14.196–14.199

Feng Q, Zhu Y, Hong J, Zhang M, Duan W, Mao N, Wu J, Xu H, Dong F, Lin F, Jin C, Wang C, Zhang J, Xie L (2014b) Growth of the large-area 2D $MoS_{2(1-x)}Se_{2x}$ semiconductor alloys. Adv Mater 26:2648–2653

Feng Q, Mao N, Wu J, Xu H, Wang C, Zhang J, Xie L (2015) Growth of $MoS_{2(1-x)}Se_{2x}$ ($x = 0.41 - 1.00$) monolayer alloys with controlled morphology by physical vapor deposition. ACS Nano 9:7450–7455

Ganatra R, Zhang Q (2014) Few-layer MoS_2: a promising layered semiconductor. ACS Nano 8:4074–4099

Geim AK, Grigorieva IV (2013) Van der Waals heterostructures. Nature 499:419–425

Gong C, Zhang H, Wang W, Colombo L, Wallace RM, Cho K (2013) Band alignment of two-dimensional transition metal dichalcogenides: application in tunnel field effect transistors. Appl Phys Lett 103:053513; (2015) Erratum. Appl Phys Lett 107:139904

Gong Y, Lin J, Wang X, Shi G, Lei S, Lin Z, Zou X, Ye G, Vajtai R, Yakobson BI, Terrones H, Terrones M, Tay BK, Lou J, Pantelides ST, Liu Z, Zhou W, Ajayan PM (2014) Vertical and in-plane heterostructures from WS_2/MoS_2 monolayers. Nat Mater 13:1135–1142

Huang J-K, Pu J, Hsu C-L, Chiu M-H, Juang Z-Y, Chang Y-H, Chang W-H, Iwasa Y, Takenobu T, Li L-J (2014) Large-area synthesis of highly crystalline WSe_2 monolayers and device applications. ACS Nano 8:923–930

Huang YL, Chen Y, Zhang W, Quek SY, Chen C-H, Li L-J, Hsu W-T, Chang W-H, Zheng YJ, Chen W, Wee ATS (2015) Bandgap tunability at single-layer molybdenum disulphide grain boundaries. Nat Commun 6:6298

Jin Y, Keum DH, An S-J, Kim J, Lee HS, Lee YH (2015) A van der Waals homojunction: ideal p-n diode behavior in $MoSe_2$. Adv Mat 27:5534–5540

Kanazawa T, Amemiya T, Ishikawa A, Upadhyaya V, Tsuruta K, Tanaka T, Miyamoto Y (2016) Few-layer HfS_2 transistors. Sci Rep 6:22277

Kang J, Tongay S, Zhou J, Li J, Wu J (2013) Band offsets and heterostructures of two-dimensional semiconductors. Appl Phys Lett 102:012111

Kang K, Xie S, Huang L, Han Y, Huang PY, Mak KF, Kim C-J, Muller D, Park J (2015) High-mobility three-atom-thick semiconducting films with wafer-scale homogeneity. Nature 520:656–660

Kaul AB (2014) Two-dimensional layered materials: Structure, properties and prospects for device applications. J Mater Res 29:348–361

Le Lay G, Aufray B, Léandri G, Oughaddou H, Biberian J-P, De Padova P, Dávila ME, Ealet B, Kara A (2009) Physics and chemistry of silicene nano-ribbons. Appl Surf Sci 256:524–529

Lee Y-H, Zhang X-Q, Zhang W, Chang M-T, Lin C-T, Chang K-D, Yu Y-C, Wang JT-W, Chang C-S, Li L-J, Lin T-W (2012) Synthesis of large-area MoS_2 atomic layers with chemical vapor deposition. Adv Mater 24:2320–2325

Lee J, Wang Z, He K, Shan J, Feng PX-L (2013) High frequency MoS_2 nanomechanical resonators. ACS Nano 7:6086–6091

Li X, Cai W, An J, Kim S, Nah J, Yang D, Piner R, Velamakanni A, Jung I, Tutuc E, Banerjee SK, Colombo L, Ruoff RS (2009) Large-area synthesis of high-quality and uniform graphene films on copper foils. Science 324:1312–1314

Li P, You Z, Cui T (2013) Molybdenum disulfide dc contact MEMS shunt switch. J Micromech Microeng 23:045026

Liang Y, Huang S, Soklaski R, Yang L (2013) Quasiparticle band-edge energy and band offsets of monolayer molybdenum and tungsten chalcogenides. Appl Phys Lett 103:0421106

Lin Y-C, Zhang W, Huang J-K, Liu K-K, Lee Y-H, Liang C-T, Chu C-W, Li L-J (2012) Wafer-scale MoS_2 thin layers prepared by MoO_3 sulfurization. Nanoscale 4:6637–6641

Lopez-Sanchez O, Llado EA, Koman V, Fontcuberta i Morral A, Radenovic A, Kis A (2014) Light generation and harvesting in a van der Waals heterostructure. ACS Nano 8:3042–3048

Lu W, Nan H, Hong J, Chen Y, Chen Z, Liang Z, Ma X, Ni Z, Jin C, Zhang Z (2014) Plasma-assisted fabrication of monolayer phosphorene and its Raman characterization. Nano Res 7:853–859

Miró P, Audiffred M, Heine T (2014) An atlas of two-dimensional materials. Chem Soc Rev 43:6537–6554

Miwa JA, Dendzik M, Grønborg SS, Bianchi M, Lauritsen JV, Hofmann P, Ulstrup S (2015) Van der Waals epitaxy of two-dimensional MoS_2-graphene heterostructures in ultrahigh vacuum. ACS Nano 9:6502–6510

Novoselov KS, Geim AK, Morozov SV, Jiang D, Zhang Y, Dubonos SV, Grigorieva IV, Firsov AA (2004) Electric field effect in atomically thin carbon films. Science 306:666–669

Ramasubramaniam A, Naveh D, Towe E (2011) Tunable bandgaps in bilayer metal dichalcogenides. Phys Rev B 84:205325

Schwierz F, Peza J, Granzner R (2015) Two-dimensional materials and their prospects for transistor electronics. Nanoscale 7:8261–8283

Tongay S, Fan W, Kang J, Park J, Koldemir U, Suh J, Narang DS, Liu K, Ji J, Sinclair R, Wu J (2014) Tuning interlayer coupling in the large-area heterostructures with CVD-grown MoS_2 and WS_2 monolayers. Nano Lett 14:3185–3190

Utama MIB, Zhang Q, Zhang J, Yuan Y, Belarre FJ, Arbiol J, Xiong Q (2013) Recent developments and future directions in the growth of nanostructures by van der Waals epitaxy. Nanoscale 9:3570–3588

Van der Zande AM, Huang PY, Chenet DA, Berkelbach TC, You Y, Lee G-H, Heinz TF, Reichman DR, Muller DA, Hone JC (2013) Grains and grain boundaries in highly crystalline monolayer molybdenum disulphide. Nat Mater 12:554–561

Vogt P, De Padova P, Quaresima C, Avila J, Frantzeskakis E, Asensio MC, Resta A, Ealet B, Le Lay G (2012) Silicene: compelling experimental evidence for graphene like two-dimensional silicon. Phys Rev Lett 108:155501

Wang S, Wang X, Warner JH (2015) All chemical vapor deposition growth of MoS_2: h-BN vertical van der Waals heterostructures. ACS Nano 9:5246–5254

Woomer AH, Farnsworth TW, Hu J, Wells RA, Donley CL, Warren SC (2015) Phosphorene: synthesis, scale-up, and quantitative optical spectroscopy. ACS Nano 9:8869–8884

Wu S, Huang C, Aivazian G, Ross JS, Cobden DH, Xu X (2013) Vapor-solid growth of high optical quality of MoS_2 monolayers with near-unity valley polarization. ACS Nano 7:2768–2772

Yu Y, Li C, Liu Y, Su L, Zhang Y, Cao L (2013) Controlled scalable synthesis of uniform, high-quality monolayer and few-layer MoS_2 films. Sci Rep 3:1866

Yu JH, Lee HR, Hong SS, Kong D, Lee H-W, Wang H, Xiong F, Wang S, Cui Y (2015) Vertical heterostructure of two-dimensional MoS_2 and WSe_2 with vertically aligned layers. Nano Lett 15:1031–1035

Yue R, Barton AT, Zhu H, Azcatl A, Pena LF, Wang J, Peng X, Lu N, Cheng L, Addou R, McDonnell S, Colombo L, Hsu JWP, Kim J, Kim MJ, Wallace RM, Hinkle CL (2015) $HfSe_2$ thin films: 2D transition metal dichalcogenides grown by molecular beam epitaxy. ACS Nano 9:474–480

Zeng Z, Yin Z, Huang X, Li H, He Q, Lu G, Boey F, Zhang H (2011) Single-layer semiconducting nanosheets: high-yield preparation and device fabrication. Angew Chem Int Ed 50:11093–11097

Zhang Y, Zhang L, Zhou C (2013) Review of chemical vapor deposition of graphene and related applications. Acc Chem Res 46:2329–2339

Zhang W, Huang Z, Zhang W, Li Y (2014) Two-dimensional semiconductors with possible high room temperature mobility. Nano Res 7:1731–1737

Zhang C, Chen Y, Huang J-K, Wu X, Li L-J, Yao W, Tersoff J, Shih C-K (2015) Visualizing band offsets and edge states in bilayer-monolayer transition metal dichalcogenides lateral heterostructures. Nat Commun 7:10349

Zhou H, Wang C, Shaw JC, Cheng R, Chen Y, Huang X, Liu Y, Weiss NO, Lin Z, Huang Y, Duan X (2015) Large area growth and electrical properties of *p*-type WSe_2 atomic layers. Nano Lett 15:709–713

Zhu W, Low T, Lee Y-H, Wang H, Farmer DB, Kong J, Xia F, Avouris Ph (2014) Electronic transport and device prospects of monolayer molybdenum disulphide grown by chemical vapour deposition. Nat Commun 5:3087

Chapter 3
Electronic Devices Based on Atomically Thin Materials

Abstract This chapter is dedicated to electronic devices based on 2D materials. The chapter starts with transistors, which are the backbone of integrated circuits, and continues with more complex devices and circuits, such as sensors, integrated circuits, and memories. This research area is in its infancy since only flakes of 2D materials are used. Thus, only few devices can be integrated, and the reproducibility and scalability issues are still not solved. The electronic devices based on 2D materials have a long path ahead before reaching the performances of Si-based integrated circuits, which actually contain up to five billions transistors/chip.

3.1 Atomically Thin Transistors

Although the 2D materials are quite new, there are already many attempts to fabricate transistors on them. These transistors are in their large majority FETs, which incorporate 2D materials as channels. Many of them contain flakes of various 2D materials, since the growth of these atomically thin materials is in infancy. A nice review about atomically thin transistors can be found in (Schwierz et al. 2015), which explains the constraints and advantages of 2D transistors. A major advantage of atomically thin transistors fabricated on 2D materials with bandgap compared to graphene FETs (GFETs), with no bandgap, is that they can attain much larger on-off ratios. Because the on-off ratio of FETs is a critical parameter for digital applications, GFETs are not considered very promising candidates for digital electronics, in contrast to FETs based on semiconducting atomically thin materials with on-off ratios typically higher than 10^4. In fact, the on-off ratio is given by

$$I_{on}/I_{off} \approx \exp[E_g/mk_BT] \tag{3.1}$$

with m of the order of unity, from where it follows that a good on-off ratio can be obtained only if

M. Dragoman and D. Dragoman, *2D Nanoelectronics*,
NanoScience and Technology, DOI 10.1007/978-3-319-48437-2_3

$$E_g > 0.4\,\text{eV} \tag{3.2}$$

Unlike the on-current, the off-current is strictly dependent on the bandgap, but the lesson learned from graphene transistors is that the lack of a bandgap, which implies a lack of a saturation region, is a detrimental fact since very often the drain conductance is larger than the transconductance and the FET is not amplifying, even at low frequencies. So, although RF transistors work in the region of on-drain current, up to now graphene RF transistors do not show significant gain beyond 5 GHz. For example, an integrated circuit containing a two-stage amplifier based on graphene FETs is working at 5 GHz with a gain of 4 dB, while the state-of-the-art is an amplifier working at 1 THz with a gain of 9 dB (Mei et al. 2015). This example shows clearly that a semiconductor with bandgap and high mobility, which can be fabricated via an advanced technology with high yield, is needed for a high-performance FET. GFETs do not satisfy these prerequisites except for the high mobility but, as discussed in the last chapter, there are methods to fabricate TMDs with high yield. The majority of TMDs have large-enough bandgaps, although the mobility is lower than in graphene or InP.

Another important requirement for any transistor is a low contact resistance. In any FET, there is a source contact resistance, R_S, and a drain contact resistance, R_D. The finite value of the source resistance impedes an efficient control of the channel, which remains ungated, since a part of the gate voltage V_G drops on the source resistance:

$$V_G = V_{Gi} + I_D R_S \tag{3.3}$$

where V_{Gi} is the intrinsic gate voltage applied directly on the FET channel. A similar situation occurs in the case of the drain resistance, the intrinsic drain voltage V_{Di} being smaller than the applied drain voltage V_D:

$$V_D = V_{Di} + I_D(R_S + R_D) \tag{3.4}$$

In general, the source and drain contact resistances are identical, the common value being denoted as contact resistance, R_c.

The contact resistance values in Fig. 3.1, extracted from several references, show that the most mature transistors have the lowest series resistances: Si FETs have resistances of 0.01 Ω mm and InP FETs of 0.03 Ω mm. GFETs with Ni and Ti electrodes have comparable series resistances, of 0.01–0.06 Ω mm, but other TMDs FETs have series resistances higher with two orders of magnitude than the state-of-the-art value. It must be pointed out that the above-mentioned extremely low resistance values for GFETs are found only in few papers, while the majority of such devices have a larger contact resistance. Considerable effort is invested nowadays to improve this situation (Allain et al. 2015), i.e. to decrease the contact resistance. However, this parameter is large in 2D materials because there is no resistive contact between metal and the 2D material. The majority of atomically thin TMD semiconductors form Schottky contact with the metals used in the

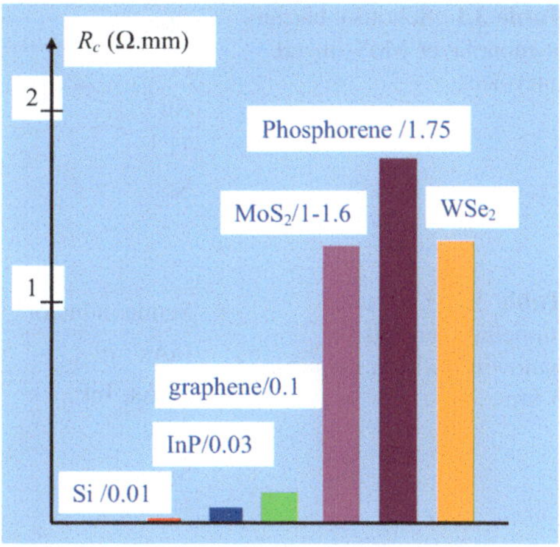

Fig. 3.1 Contact resistance of FETs based on various semiconductors

semiconductor cleanroom technology. In particular, the most studied TMD semiconductor—MoS_2—forms Schottky barriers with the most common metals, the values of which are given in Table 3.1 (please note that the Schottky barriers, as well as many other parameters of 2D materials, may have different values in different references). The contact resistance depends also on the number of layers, doping and growth conditions of 2D materials.

There are many other important aspects for 2D transistors which need to be properly addressed. For instance, CMOS FET digital electronics requires FETs with semiconductor channels with high mobility, high saturation velocity and low effective mass. However, if the drain-source distance becomes lower than 5 nm, the tunneling current between source and drain cannot be suppressed in existing high-mobility and low-effective mass semiconductors used today. Then, 2D materials with an effective mass larger than 0.5 m_0 can replace them.

In this respect, a simple but effective method can be used to estimate the undesired tunneling currents (Schwierz et al. 2015). By using the following analytical expression for the transmission between source and drain of nanowire transistors with a barrier of width W = 5 nm (Sylvia et al. 2012):

$$T_{SD} = \left(1 + \frac{V_0^2 \sinh^2(|k|W)}{4E(V_0 - E)}\right)^{-1} \quad (3.5)$$

where V_0 is the barrier height, and

$$k = [2m_{eff}(E - V_0)]^{1/2}/\hbar \quad (3.6)$$

is the electron wavenumber, the tunneling current is given by

Table 3.1 Schottky barriers at monolayer MoS_2/metal interfaces

Metal	Schottky barrier (eV)
Pt	0.25
Au	0.2
Ti	0.06
Sc	0.03

Table 3.2 Unwanted tunneling currents in nanowires with a diameter of 5 nm

Semiconductor	Tunneling current (μA)
InAs	0.2
GaAs, InP	2×10^{-2}
GaN	7×10^{-4}
Si	10^{-4}

$$I \propto M \times T_{SD} \tag{3.7}$$

In (3.7) M is the number of open channels. The tunneling currents in nanowires of 5 nm in diameter are displayed in Table 3.2.

At a gate length of 5 nm, the direct tunneling current between source and drain is the main leakage current source for both FETs and tunneling FETs. As can be seen from Table 3.2, Si is still the best choice for nanowire transistors at the 5 nm scale; the leakage current is lower for materials with higher effective mass of charge carriers.

A 2D material FET transistor with a 5 nm gate length is represented in Fig. 3.2. Such a transistor was analyzed numerically by (Cao et al. 2015), and it was found that bilayer MoS_2 FETs can have high performances up to gate lengths of 6.6 nm, while WSe_2 FETs outperform MoS_2 FETs for gate lengths up to 5.9 nm, thus satisfying the ITRS map up to 2026. This is a promising result, since the research dedicated to FETs based on 2D materials is only at the beginning.

The first MoS_2 monolayer transistor was reported by (Radisavljevic et al. 2011). The channel was a flake of MoS_2 monolayer, the transistor having a top gate isolated by HfO_2 from the channel and a doped Si backgate isolated from channel by a SiO_2 layer. The structure of this transistor, with a length $L = 1.5$ μm and a width $W = 4$ μm, is presented in Fig. 3.3. The thicknesses of the HfO_2 and SiO_2 layers are 30 nm and 270 nm, respectively. The mobility, determined from current-voltage dependences, was found to have a high value, of about 200 cm^2/V s, due to the HfO_2 high-permittivity dielectric layer, with $\varepsilon_r = 25$, which suppresses the Coulomb scattering. The mobility is similar to that of a 10 nm-wide graphene nanoribbon with a bandgap of 0.4 eV, although the bandgap of the MoS_2 monolayer is four times larger. The on-off ratio of the transistor in Fig. 3.2 is higher than 10^8 for relatively low drain voltages, in the 0.01–0.5 V range, while the drain current is rather weak, not exceeding 1 μA for a backgate voltage of 40 V. The transconductance of the transistor, of 4 μS (or 1 μS/μm), is rather modest, but comparable to that of high-performance graphene transistors, which reach values

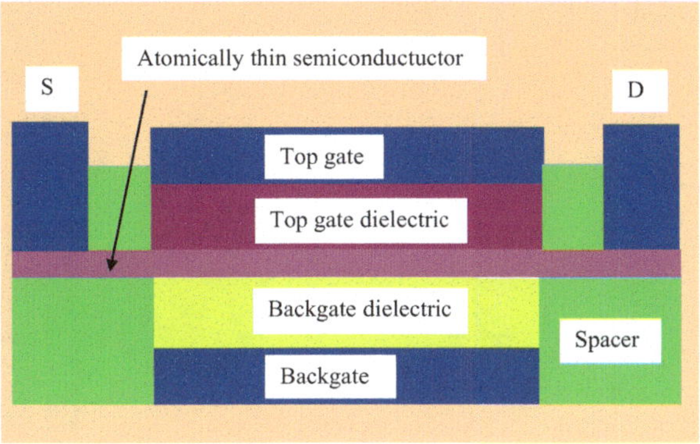

Fig. 3.2 A 5 nm-gate-length FET with 2D material channel

higher than 1.2 μS/μm. The absence of a high transconductance is due to the fact that the I_D-V_D dependence at various top gates ranging from −2.5 V up to +4 V is linear, like in graphene, and shows no saturation region. The I_D-V_{GT} dependence is constant in the top-gate voltage range −2 V to +4 V, which suggests that the channel is controlled by the top gate only for top gate voltages in the range −4 V to −2 V. This rather reduced gate swing determines a subthreshold slope of 74 mV/decade. Despite these modest performances, this first attempt of 2D materials applications in nanoelectronics triggered many subsequent research efforts.

A high-performance MoS_2 FET was fabricated using scandium contacts, because the difference between the electron affinity of MoS_2 (4 eV) and the workfunction of Sc is quite small (ϕ_{Sc} = 3.5 eV), and the contact resistance is much lower (Das et al. 2013). After extracting the Schottky barrier heights between several metals and MoS_2 (Ni: 150 meV, Pt: 230 meV, Ti: 50 meV and Sc: 30 meV), it was determined that the MoS_2-metal contact is dominated by Fermi level pinning near the conduction band of MoS_2. The transistor was again

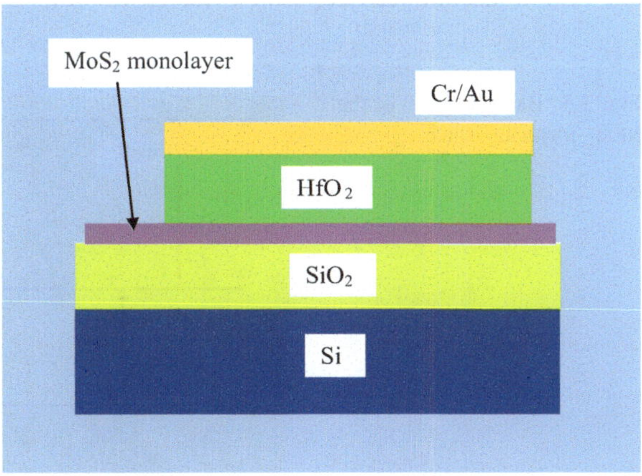

Fig. 3.3 MoS_2 monolayer transistor

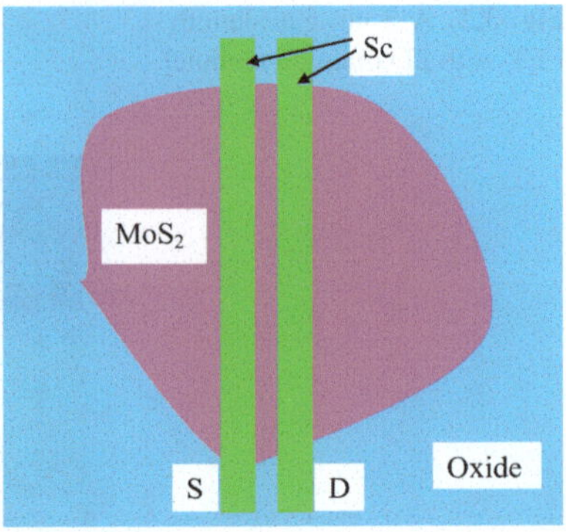

Fig. 3.4 MoS_2 FET with Sc contacts

fabricated on MoS_2 flakes with a thickness of 2–12 nm, as shown in Fig. 3.4, in a backgate configuration, and had a channel length of 5 μm.

The band diagram of MoS_2-metal interface is presented in Fig. 3.5. Electron injection for n-MoS_2 occurs only for Sc and Ti contacts, while Ni and Pt should be used for hole injection. However, irrespective of the metal type, the MoS_2 FET is of n-type, indicating that the Fermi level is located near the conduction band. A degradation of the on-state for positive $V_G - V_{th}$ is evidenced for the metals in Fig. 3.5, from Sc up to Pt, and can be explained by tunneling through the corresponding Schottky barriers at the source-channel interface. This phenomenon has as overall effect the limitation of current injection in the on-state and in the off-state in the subthreshold region. Note that V_{th} denotes the threshold voltage, which is the minimum gate voltage for which a conducting channel between drain and source is created; below V_{th}, in the subthreshold regime, the drain current depends exponentially on the gate voltage.

The type of the metal contact influences dramatically the MoS_2 FET performances, as can be seen from Fig. 3.6. This figure indicates that a smaller Schottky barrier height produces a significantly higher current injection, which is reflected in the FET transconductance, which decreases from 0.28 μS/μm for Sc contacts to

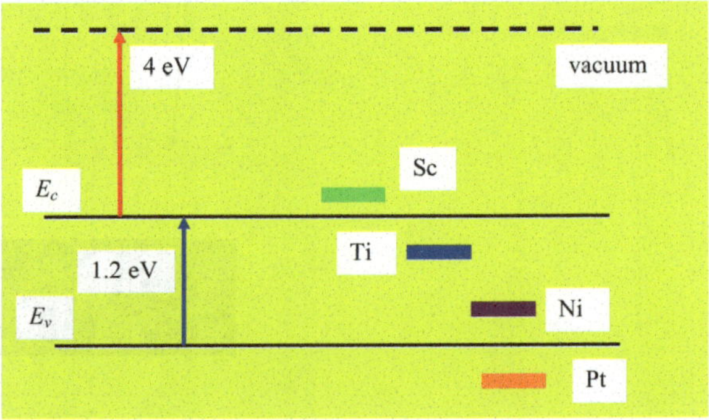

Fig. 3.5 Band diagram of the MoS_2-metal interface

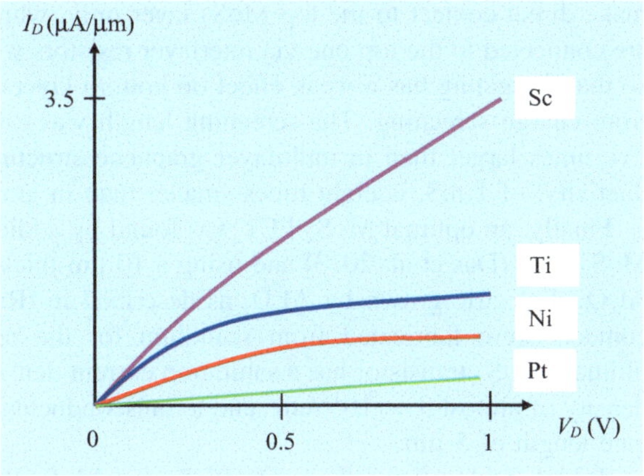

Fig. 3.6 Drain current-drain voltage dependence in a 10-nm-thick MoS_2 FET at $V_G - V_{th} = 4$ V (after Das et al. 2013)

Table 3.3 Mobility of MoS_2 as a function of metal contact type

Metal	MoS_2 mobility (cm^2/V s)
Pt	21
Ni	36
Ti	125
Sc	184

only 0.03 μS/μm for Pt contacts. The mobility for different metal contacts is displayed in Table 3.3.

The I_D-V_D characteristics depend also strongly on the number of MoS_2 monolayers (see Fig. 3.7). The drain current reaches a maximum value for a thickness of 10 nm, while for thicker (20 nm) or thinner flakes (6–2 nm), the drain currents are at least three times lower. According to (Das et al. 2013), the existence of an optimal layer thickness for high-performance MoS_2 FETs in the range 6–12 nm is related to two contradictory aspects: (i) the metals for drain an source electrodes

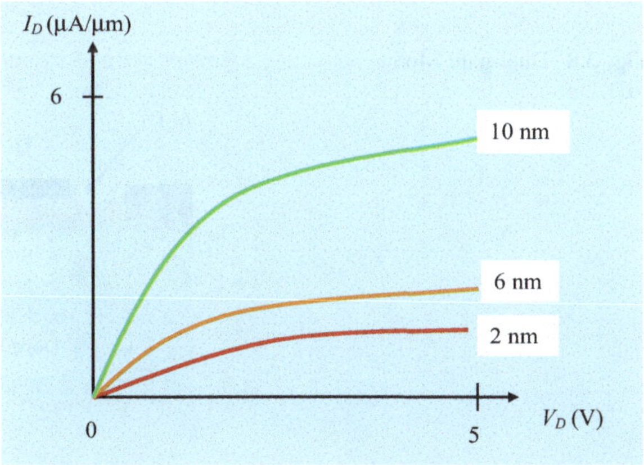

Fig. 3.7 Drain current-drain voltage dependence on the thickness of the MoS_2 FET at $V_G - V_{th} = 2$ V (after Das et al. 2013)

make direct contact to the top MoS_2 layer only, while the additional MoS_2 layers are connected to the top one via interlayer resistors with a finite conductivity σ_{inter}, so that the gating has a weak effect on bottom layers, and (ii) the top layers suffer from charge screening. The screening length was estimated to be 3 nm, which is five times larger than in multilayer graphene structures, while the interlayer conductivity, of 1 mS, is eight times smaller than in graphene multilayers.

Finally, an optimal MoS_2 FET was found by adding a top gate to the backgated MoS_2 FET (Das et al. 2013) and using a 10 nm-thick MoS_2 channel with a 15 nm Al_2O_3 dielectric grown by ALD, as described in (Radisavljevic et al. 2011). The contacts were fabricated from scandium for the reason explained above. This ultimate MoS_2 transistor has a saturation current density of 240 μA/μm for a carrier density of almost $3 \times 10^{12}/\mathrm{cm}^2$ and a transconductance of 4.7 μS/μm for the same gate length of 5 μm.

A high-performance dual-gate FET on a MoS_2 flake, as displayed in Fig. 3.8, was fabricated and studied by (Liu and Ye 2012). This configuration is the same as in (Das et al. 2013), i.e. it has both top and back gates, the contacts being 20/50 nm Ni/Au for the source and drain, and Ti/Au for the gate. The MoS_2 channel is 15 nm thick and has a bandgap of 1.2 eV, the drain current density being about 3 mA/mm at a backgate voltage of 40 V and 10^{-2} mA/mm for a top gate voltage of 4 V. The on-off ratio reaches 10^8 and the top-gate transconductance is 0.61 mS/mm, but the corresponding mobility is weak, of only 4.13 $\mathrm{cm}^2/\mathrm{V\,s}$, due to the contact and access resistance, more precisely due to a very large Schottky barrier.

Multilayer MoS_2 transistors are high-performance thin film transistors (TFTs) (Kim et al. 2012), which exhibit a high mobility (>100 $\mathrm{cm}^2/\mathrm{V\,s}$) and a low, almost ideal subthreshold slope, of 70 mV/decade at room temperature. MoS_2 TFTs can become alternative devices for high-resolution liquid crystal displays and organic light emitting diodes (OLEDs), being able to enhance the current by generating multiple conduction channels via the field effect, as in MOSFETs. The MoS_2 TFT in (Kim et al. 2012), fabricated on MoS_2 multilayer with a thickness of 30 nm, had an Al_2O_3 backgate insulator, with a thickness of 50 nm, deposited via ALD. Moreover, it showed saturation regions over a large range of voltages.

Fig. 3.8 Dual-gate MoS_2 FET

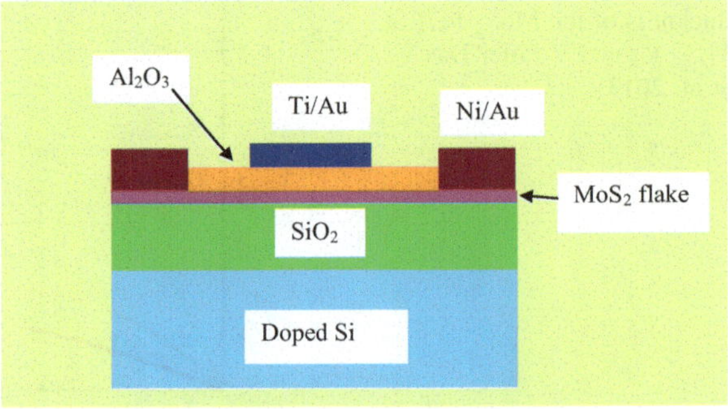

The subthreshold slope is given by

$$SS = [1 + (C_S + C_{it})/C_{ox}](k_B T/e) \ln 10 \tag{3.8}$$

where C_S is the channel capacitance, which is generally negligible,

$$C_{it} = eD_{it} \tag{3.9}$$

is the capacitance of interface traps with D_{it} the corresponding density of traps, and $C_{ox} = \varepsilon_{ox}/t_{ox}$ is the capacitance of the oxide with thickness and dielectric permittivity t_{ox} and ε_{ox}. For a trap density of 2.6 × 10^{11} $eV^{-1}cm^{-2}$, *SS* = 24 mV/decade at 77 K and 70 mV/decade at room temperature. How is it possible to obtain such a low *SS* value compared to the state-of-the-art Si transistors, considering that efforts of lowering the *SS* in CMOS transistors have been ongoing for a couple of decades? The reason is that the density of defects at the Si/dielectric interface is much higher than at the MoS_2/dielectric interface, because in layered 2D materials there are no out-of-plane broken bonds. As a result, the traps are located in the dielectric, but not in the semiconductor (Kim et al. 2012). This amazing property of layered 2D materials compared to Si plays an important role, since the subthreshold region is a major contributor in power consumption. The MoS_2 TFT transistor is depicted in Fig. 3.9.

Highly bendable MoS_2 FETs with high-permittivity dielectrics are also reported (Chang et al. 2013). The flexible FET has a subthreshold slope *SS* = 82 mV/decade, a high mobility of 30 cm^2/V s, and preserves its characteristics up to a bending radius of 1 mm. The transistors were made with flakes having thicknesses in the range 7.9–23.5 nm. Their on-off ratio is 10^7 and the intrinsic gain g_m/g_d is higher than 100. All these parameters indicate an amplifying flexible transistor, with an amazing *SS* value, which recommends it for low-power applications. The flexible

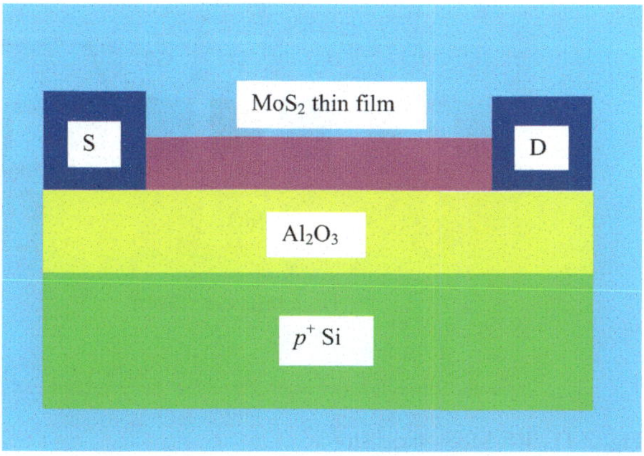

Fig. 3.9 Schematic representation of the MoS_2 TFT

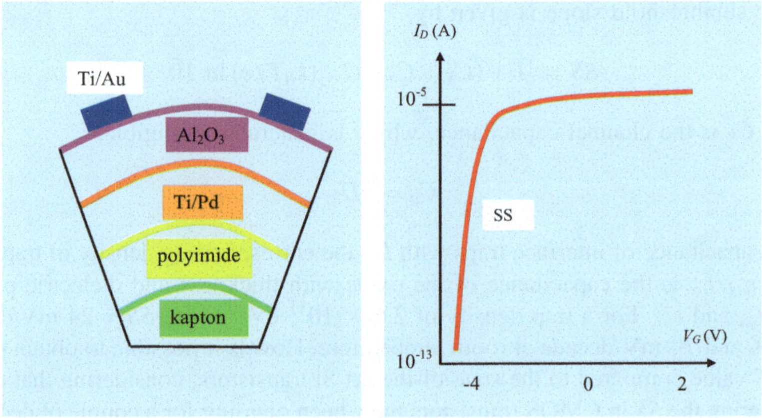

Fig. 3.10 Bendable MoS_2 FET and its I_D-V_G characteristic

FET is illustrated in Fig. 3.10. The high subthreshold slope value indicates that there is a low leakage current. The *SS* parameter can be expressed also as

$$SS = (\partial \log I_D / \partial V_G)^{-1} \quad (3.10)$$

MoS_2 FETs operate also at GHz frequencies and have reached already a cutoff frequency of 6 GHz (Krasnozhon et al. 2014). The double-gate transistors working in RF were fabricated using MoS_2 flakes with thicknesses varying from a single monolayer up to a multilayer (5 nm). The cross-section of the RF transistor, shown in Fig. 3.11, indicates a coplanar configuration with a double grounded source, a central electrode, which is the drain contact, and two gates, G1 and G2. The source and drain electrodes are from Au, the gate electrodes are Cr/Au and the dielectric is HfO_2, deposited via ALD, with a thickness of 30 nm. The on-state current for a MoS_2 monolayer channel attains 300 μA/μm for a drain voltage of 2 V. However,

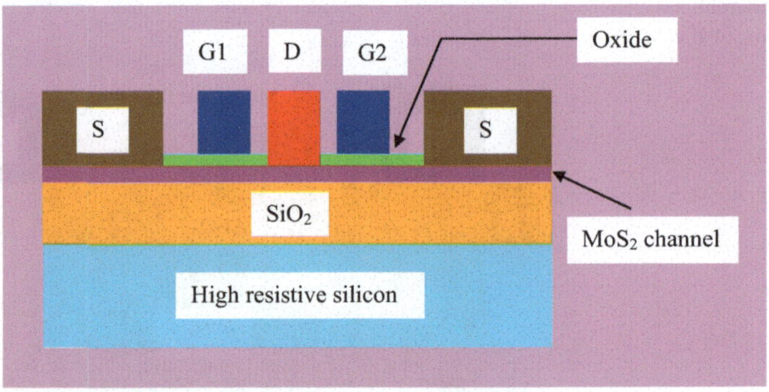

Fig. 3.11 RF MoS_2 transistor

in this case the mobility is moderate, of 85 cm^2/V s, and the contact resistance $R_c = 2$ kΩ/μm is large. The transconductance reaches about 50 μS/μm at a drain voltage of 1.5 V, and the maximum oscillation frequency is $f_{max} = 2.2$ GHz for a device with a gate length of 240 nm. For a FET with a MoS_2 monolayer the voltage gain is about 1.8 at 1 GHz, whereas for a transistor with trilayer MoS_2 this parameter becomes 1.5 at 5 GHz.

RF transistors and circuits were fabricated also on MoS_2 grown by CVD over a large area of few mm^2 (Sanne et al. 2015). Due to the high-quality growth of MoS_2, the MoS_2 FETs display a current density higher than 200 μA/μm, a maximum transconductance of 38 μS/μm, a cutoff frequency $f_T = 6.7$ GHz, and a maximum frequency of 5.3 GHz for a gate length of 250 nm. The amplification is 6 dB at 100 MHz, voltage amplification occurring up to 3 GHz. A common source amplifier with a gain of 5 (14 dB) and a mixer with a conversion of −15 dB both working at MHz frequencies were fabricated based on this MoS_2 FET.

As for graphene transistors, MoS_2 FETs using self-aligned gates attain the best performances (Cheng et al. 2014). In this case, the substrate was doped Si/SiO_2 and MoS_2 flakes were transferred on it. Then, metal gates wrapped in dielectric (Al_2O_3) were transferred on the substrate (see Fig. 3.12). The e-beam lithography (EBL) method and metal deposition was used to define first the external source, drain and gate Ti/Au (50 nm/50 nm) electrodes and then the Ni/Au (5 nm/10 nm)

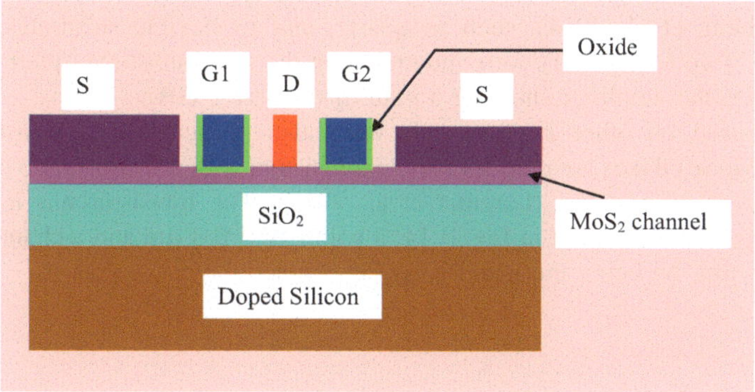

Fig. 3.12 Cross-section view of the self-aligned gate MoS_2 FET

Fig. 3.13 Top view of the coplanar electrode for MoS_2 self-aligned FET

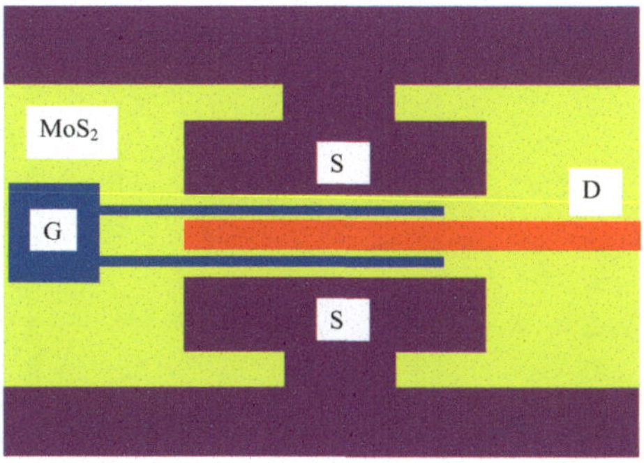

self-aligned source and drain electrodes, which have a minimum access resistance and parasitic capacitance. The transistor, displayed in Fig. 3.13, is fabricated with coplanar electrodes, as in similar graphene self-aligned transistors. The source electrodes are grounded, the gate electrode is the input signal electrode, while the output signal electrode is the drain. The channel has a length of 1 μm and a width of 4 μm, and the backgate capacitance is 11.5 nF/cm^2. The mobility reaches 170 cm^2/V s for a drain-voltage of 0.5 V and a backgate voltage of 20 V.

The MoS_2 thickness is a critical parameter for high-performance MoS_2 FETs. If the thickness of the MoS_2 multilayer is higher than 20 nm, there is significant screening at the bottom electrode and the FET cannot be turned on or off. The suitable thickness was found to be in the range 2–7 nm (Cheng et al. 2014). By reducing the dimensions of the channel length at 116 nm, the self-aligned gate MoS_2 FET considered above reached an improved transconductance of 60 μS/μm at a drain voltage of 5 V, and a drain conductance of less than 2 μS/μm, indicating an intrinsic gain of 30 and an on-off ratio higher than 10^7. This is the best high-frequency MoS_2 FET reported up to now, since the cutoff frequency is $f_T =$ 42 GHz and $f_{\max} = 50$ GHz. From these data, the saturation velocity in MoS_2 is determined as

$$v_{sat} = 2\pi L f_T = 1.8 \times 10^6 \,\text{cm/s} \tag{3.11}$$

The search for high-performance MoS_2 transistors was extended to other substrates (Cheng et al. 2014), such as quartz, and to flexible substrates, such as polyimide. Two MoS_2 FETs were integrated on the same substrate (quartz) to form an amplifier, the amplifier showing a gain of 6 dB at 1 GHz.

FETs based on other 2D materials were also fabricated and measured. For example, $MoSe_2$ flakes are used for n-type transistors with an on-off ratio of 10^6 and a mobility of 50 cm^2/V s (Larentis et al. 2012). The transistor has a backgate configuration, as illustrated in Fig. 3.14, a width $W = 0.8$ μm and a channel length of $L = 1.8$ μm, Ni being the metal used for contacts.

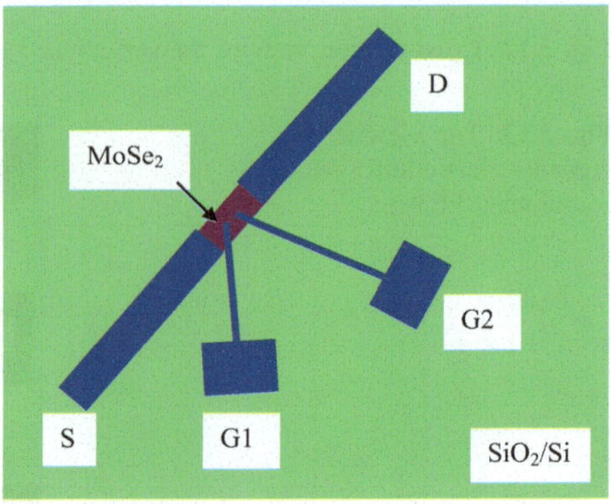

Fig. 3.14 The $MoSe_2$ FET

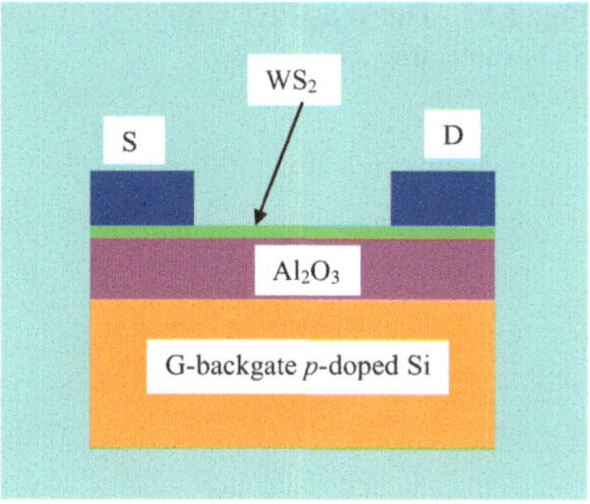

Fig. 3.15 The WS_2 FET with HfO_2 dielectric

The WS_2 FET (Hwang et al. 2012) is an ambipolar transistor, with an on-off ratio higher than 10^5 and a clear saturation region. The transistor, depicted in Fig. 3.15, is fabricated using WS_2 flakes with a thickness of 13 nm, i.e. about 18 monolayers. More precisely, the WS_2 flakes are directly deposited on the Al_2O_3 layer, in turn deposited previously by ALD on a *p*-doped Si substrate. The source and drain electrodes, defined by EBL, are of Ti (5 nm)/Au (100 nm). The channel length is 2.5 μm and its width is 2 μm. The on-off ratio attains 10^5 for $V_D = 1$ V and 10^4 for $V_D = 5$ V. The I_D-V_G characteristic of this ambipolar transistor displays a "V" dependence, as in the case of graphene, because the Fermi level of the metal is situated almost in the middle of the bandgap of WS_2 (see Fig. 3.16).

A similar WS_2 FET was reported by (Liu et al. 2014b). The FET has the same backgate configuration, as shown in Fig. 3.17, except that the dielectric is now SiO_2. The transistor still displays an ambipolar transport with a very high on-off ratio of 10^8 and a maximum mobility of 234 cm^2/V s, but the saturation region is not visible.

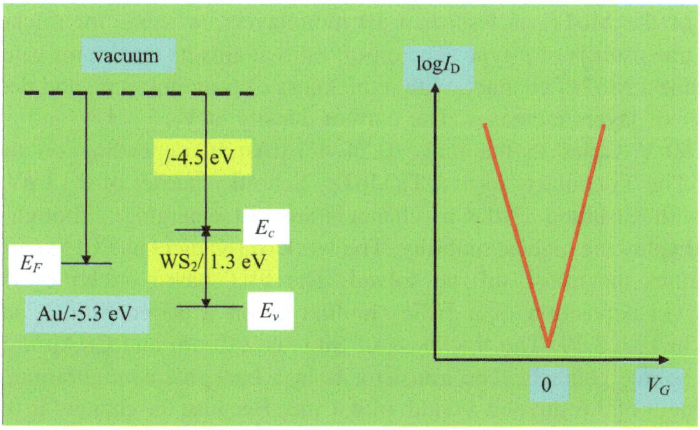

Fig. 3.16 Band diagram showing Au/WS_2 Schottky barrier formation and ambipolar transport in the WS_2 FET

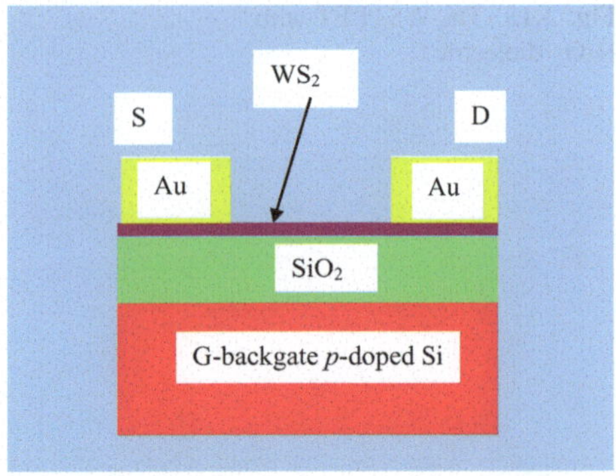

Fig. 3.17 The WS_2 FET with SiO_2 dielectric

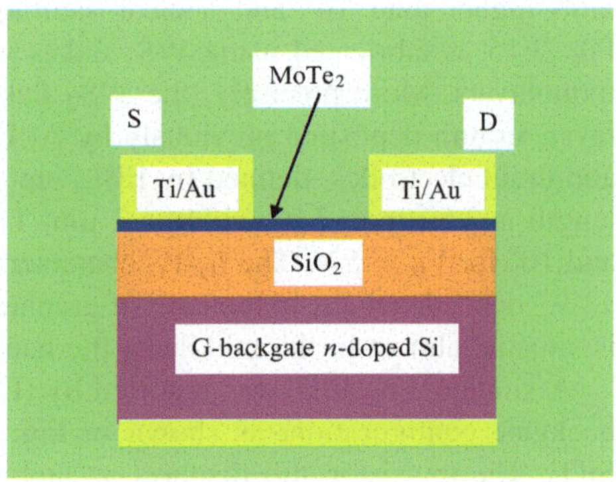

Fig. 3.18 $MoTe_2$ FET with a SiO_2 dielectric

Exfoliated multilayer $MoTe_2$ FETs with SiO_2 dielectric were also reported (see Fig. 3.18), the thickness of the multilayer $MoTe_2$ channel being in the range 6–44 monolayers (Fathipour et al. 2014). This transistor shows ambipolar transport if the thickness of the $MoTe_2$ is less than 10 monolayers, whereas for thicker $MoTe_2$ layers the transistor is of *p*-type. The on-off ratio attains its maximum value, of 10^5, for the thinnest $MoTe_2$ channel, with a thickness of 6 monolayers, and decreases as the number of layers increases. The current density at $V_D = -1$ V and a backgate voltage −80 V varies in the range 0.24–4.4 μA/μm, depending on the $MoTe_2$ thickness. The Ti contacts form a Ti/$MoTe_2$ Schottky barrier of 0.13 eV.

FETs with Hf-based TMDs as channels are just emerging, although these 2D materials display the highest mobility. The work of (Kang et al. 2015) demonstrates the difficulties that must still be solved. The FET based on $HfSe_2$ flakes was fabricated via deposition of a $HfSe_2$ multilayer on a doped Si/SiO_2 surface, as displayed in Fig. 3.19. The thickness of SiO_2 is 285 nm and Cr/Au are used for drain and source contacts. The transistor is in a backgate configuration, having a channel length of 3.6 μm and a width of 4.4 μm. Because the channel is sensitive to ambient conditions, PMMA was used for passivating the entire flake of $HfSe_2$ with a thickness of 17.3 nm, i.e. 28 layers. The FET is of *n*-type, with a threshold voltage

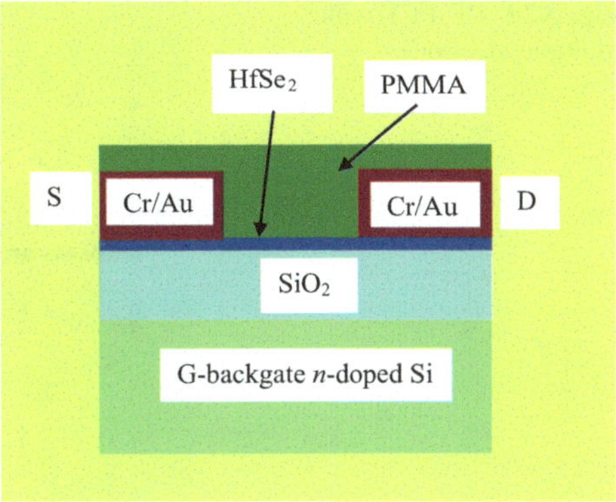

Fig. 3.19 The $HfSe_2$ FET with SiO_2 dielectric

of 24 V. At negative backgate voltages, the electrons are depleted and the drain current is 8 pA, while at positive backgate voltages the drain current reaches 0.05 μA. The drain current-drain voltage dependence shows no saturation region, and the on-off ratio is 0.5×10^5. The mobility varies with temperature as $\mu \propto T^{-1.99}$, and is about 0.22 cm^2/V s at 280 K. This value must be compared with the theoretical one given in the previous chapter, of 3579 cm^2/V s, to understand that the differences between ab initio calculations and reality/experiments are huge in TMDs, and new type of contacts and passivation techniques are necessary to be investigated in order to improve the performances of FETs made from these materials.

Black phosphorous (BP) is a layered material, like graphite, with monolayers stacked together due to vdW forces. Phosphorene (PH) is the name of BP monolayer, just as graphene is the name of a monolayer of graphite. PH is semiconductor, with a direct bandgap of 2 eV, while few layers of BP have a band gap of 0.2–0.3 eV verified by ARPES by (Li et al. 2014). Because BP is known to have higher mobilities than MoS_2, it is expected to behave as a graphene-like 2D material with bandgap. The majority of applications of BP is based on few-layer flakes, which are stable in air, while PH is very sensitive to ambient conditions and needs passivation.

An ambipolar BP FET with a backgate configuration and a mobility of 1000 cm^2/V s has been demonstrated by (Li et al. 2014). The back gate is doped Si, over which 90 nm SiO_2 is grown. Flakes of different thickness, up to 50 nm, were deposited on Si/SiO_2. Cr/Au (5 nm/60 nm) contacts are further deposited via e-beam evaporation and all measurements are performed in high-vacuum. The FET with a 5-nm-thick BP channel has an on-of ratio of 10^5. At a certain ratio between the channel length, which is 4.5 μm in the device (the channel width is 2.3 μm), and the SiO_2 thickness, saturation regions are observed in the *p*-type FET. Otherwise, the transport is ambipolar and no saturation occurs due to the limitations of doping by the breakdown field of SiO_2. The subthreshold slope *SS* is very large, exceeding 3 V/decade, because the dielectric is too thick. The transistor has a hole mobility of 984 cm^2/V s for a 10 nm thick BP, which is similar to that in Si devices. The mobility depends strongly on the substrate thickness, the highest

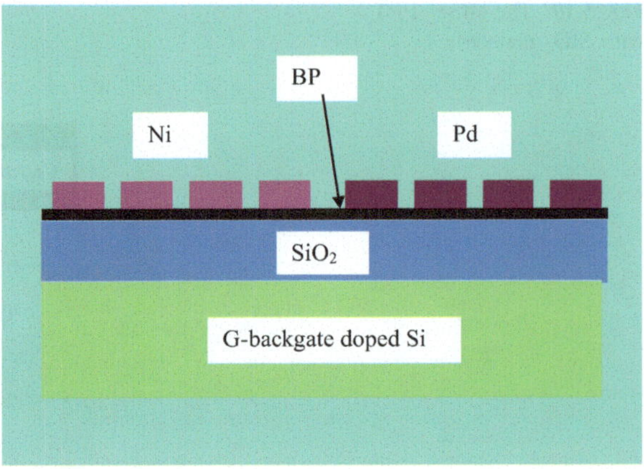

Fig. 3.20 BP FET with multiple electrodes

values being obtained for a 10-nm-thick substrate. Thin BP layers have reduced mobilities due to charge impurities located at the interface, while in thicker samples the induced charges are screened. Thus, if the thickness of BP increases at 20 nm, the mobility becomes 200 cm^2/V s. The mobility decreases slowly with temperature, with a factor $T^{-0.5}$.

In search for better BP-metal contacts, Ni and Pd contacts were deposited on few-layer BP, in turn deposited on a Si/SiO_2 substrate as before, which plays the role of the backgate (Du et al. 2014). The BP FET is displayed in Fig. 3.20. Again, for Ni contacts ambipolar transport was observed in the drain current-backgate voltage dependence, which has the characteristic V-like imprint. The measured extrinsic mobility is 170 cm^2/V s, the on-off ratio is only 10^2, and the on-drain current attains 93.3 mA/mm for a channel length of 3 μm and a BP thickness of 18.7 nm. In a BP FET with the same geometry but with Pd contacts, the on-drain current is 111.8 mA/mm and the mobility reaches 186.5 cm^2/V s. The BP FET with Pd contacts shows a less pronounced ambipolar transport compared to that with Ni contacts due to the higher workfunction of Pd compared to Ni. The drain current depends strongly on the channel length. For instance, in the device with Ni contacts, the drain current in the *n*-doped regime (for positive gate voltages) is one order of magnitude higher for a channel length of 100 nm, than for a channel length of 2 μm. A similar behavior is encountered in the case of Pd contacts. The mobility is higher for longer channel lengths (up to 3 μm), decreasing with about two orders of magnitude for smaller lengths, of 10 nm.

BP FETs, as any TMD-based FETs, are competing for the best sub-5 nm transistors. In comparison to TMDs, the mobility of BP is higher, attaining 10^3 cm^2/V s for FET channels with few-μm length. The suppression of short channel effects requires mainly:

- a thin channel; the thickness of the channel must be less than one third of the gate length to assure full electrostatic control of the gate on the channel concentration of carriers

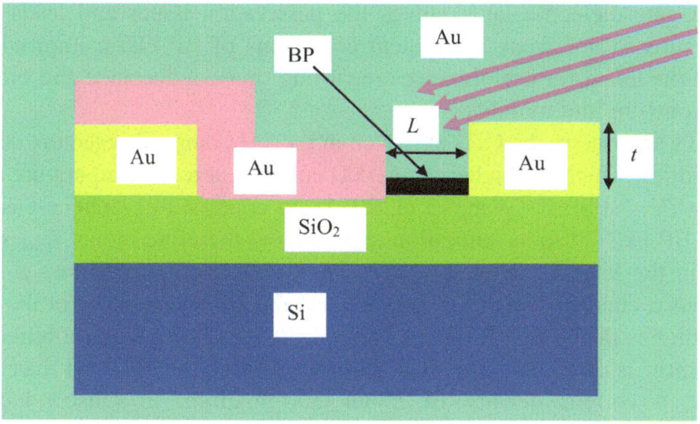

Fig. 3.21 Formation of an ultrashort channel via tilted deposition

- reduced roughness/scattering; few atomic layers of conventional semiconductors are very rough, and thus the mobility decreases dramatically due to scattering, while 2D atomically thin layered semiconductors such as TMDs or BP are smooth and without dangling bonds on the surface.

These important issues were recently tackled in a top-gate BP FET, where the Al_2O_3 dielectric, deposited by ALD with a thickness of 10 nm, plays the role of the passivation layer for the few-layer BP channel (Miao et al. 2015). The fabrication process of this device is innovative. First, a long channel, with L = 1 μm, is fabricated using flakes of BP thinner than 10 nm transferred on a doped Si wafer, which is covered with 300 nm SiO_2. 50 nm-thick Au contacts were fabricated by EBL patterning, followed by metal evaporation and liftoff. Then, a narrow channel forms in the next step, which consists of an EBL process for opening a window in the channel region, followed by 20 nm deposition of Au, evaporated at a certain angle (see Fig. 3.21). The length of the ultrashort channel is given by

$$L = t \times \tan(\theta) \tag{3.12}$$

where t is the thickness of the first electrode and θ is the evaporation angle. This p-type BP FET with an ultrashort channel length of 20 nm has an on-current of 174 μA/μm and an on-off ratio of 10^2 at a drain voltage of only 0.1 V. The on-off ratio increases with the channel length, reaching 10^4 for L = 2 μm.

BP is a material that is difficult to be processed. In particular, (Miao et al. 2015) have pointed out that the samples were passivated with PMMA between different steps of fabrication, and were stored in a glovebox filled with argon when not in use. A transconductance of 70 μS/μm and a low mobility, of 17 cm^2/V s, were found at a gate voltage of −2 V. This mobility value, lower than in other reported BP FETs, is attributed to the top gate dielectric, which increases the scattering that, combined with the short channel length, leads to velocity saturation. Thus, the

Al_2O_3 top gate dielectric layer solves the passivation issues and leads to reproducible results over 1 week in ambient conditions of BP FETs, improving at the same time the transistor performances regarding transconductance and on-off ratio, but compromising the mobility.

BP FETs are used in the GHz and even in the THz range as detectors due to their high mobilities, which are higher than 1000 cm^2/V s at room temperature and reach 50,000 cm^2/V s at 30 K (Wang et al. 2014). In contrast to graphene-based FETs, few-layer BP FETs display saturation regions, and therefore higher f_{max} values are expected in the latter devices. The FET has CPW electrodes, shows *p*-type conduction, has a channel length of 300 nm, and the BP thickness is in the range 6–10 nm, which is the best choice for a high on-off ratio and a high mobility. The top gate dielectric is HfO_2, with a thickness of 21 nm. The BP FET has Ti/Pd/Au electrodes, a current density of about 300 mA/mm, a transconductance of 180 mS/mm, an on-off ratio of 2000, and a mobility of 400 cm^2/V s. The cutoff frequency is 12 GHz and f_{max} = 20 GHz. The saturation region of this BP FET is the origin of the low drain conductance and high transconductance. The transistor has gain up to 50 GHz, the gain being 20 dB at 2 GHz and 2 dB at 10 GHz, surpassing thus years of research in RF graphene transistors, where the saturation region is missing due to the absence of a bandgap.

THz detection based on FETs was also performed using few-layer BP as channel, as was done in FETs with graphene channels (Viti et al. 2016). As in many other papers, a bulk BP crystal was grown via CVD and the famous adhesive tape method was used to exfoliate a flake of BP with thickness in the range 8–14 nm and then transfer it on a 300 nm SiO_2 grown on intrinsic Si. Two flakes of BP were selected as FET's channel, with thicknesses of 9 and 14 nm, and with different orientations. The FET on the thicker flake has a channel length of 900 nm, a gate length of 450 nm, and the source and drain electrodes were patterned by EBL in the shape of a bow tie antenna (see Fig. 3.22). The BP FET has better performances

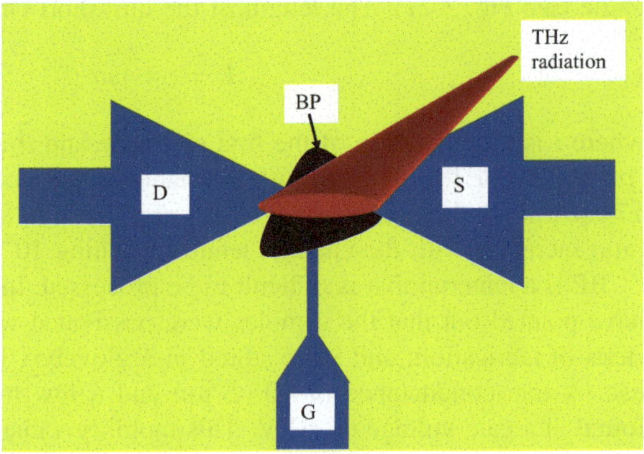

Fig. 3.22 The THz detection with few layers BP FET

than GFETs, in particular a transconductance of 11–19 nA/V, and a subthreshold slope in the range 250–1136 mV/decade, but the mobility is low, of 330 cm^2/V s.

The photoresponse of this device is given by:

$$\Delta V = -(1/\sigma)\left(\frac{d\sigma}{dV_G}\right)\left(\frac{R_L}{(1/\sigma)+R_L}\right)\eta \tag{3.13}$$

where σ is the channel conductivity, R_L is the load impedance, and η is the antenna coupling efficiency. The responsivity attains 5–8 V/W, and the signal-to-noise ratio is 500 at 0.3 THz.

FETs were made even on silicene, but they are not stable in air. For instance, the electrical signal of a silicene FET is lost in air after 2 min due to amorphous degradation of this material. A method to overcome this drawback is to grow epitaxially silicene on an Ag(111) film grown on a mica substrate, and to encapsulate silicene with a 5 nm Al_2O_3 film, followed by detaching the Ag substrate and attaching silicene on a 90 nm SiO_2 dielectric grown on a p^{++} Si substrate (Tao et al. 2015). The silicene-based FET with silver drain and source contacts has ambipolar transport and a mobility of 100 cm^2/V s at room temperature. The silicene has a bandgap of 0.21 eV that determines an on-off ratio higher than 10, which is higher than in graphene (on-off ratio 5). However, the drain current-drain voltage characteristic is linear, while the drain current-gate voltage dependence has the V-like shape typical for ambipolar transport, which indicates a Dirac point near 0 V. The silicene cannot compete with graphene or TMDs because it cannot be grown at wafer level and is very instable in air.

3.2 Atomically Thin Devices and Circuits

Because vertical graphene/BN/graphene structures have shown resonant tunneling behavior, efforts to extend this phenomenon to other systems resulted in the study of the vertical graphene/WS_2/graphene heterostructure (Georgiou et al. 2012). The combined effect of tunneling through the WS_2 barrier and thermionic transport over the barrier produces an on-off ratio of 10^6. The semiconducting WS_2 has a much lower bandgap than the insulating BN layer, and the changes in the Fermi level induced by a bottom graphene electrode acting as backgate is significant, i.e. of the order of the barrier height, so that the on-off ratio increases dramatically compared to the case when BN is used as barrier. The structure of the backgate transistor is depicted in Fig. 3.23.

The main property of this device is the subthreshold slope, which is lower than 60 mV/decade when the tunneling current dominates the total current of the heterostructure (for barrier heights > 0.4 eV). Thus, although the current is weak, low-power applications are anticipated for these vertical heterostructure transistors, which are, in principle, tunnel diodes with current-voltage dependences tunable as a function of the gate voltage.

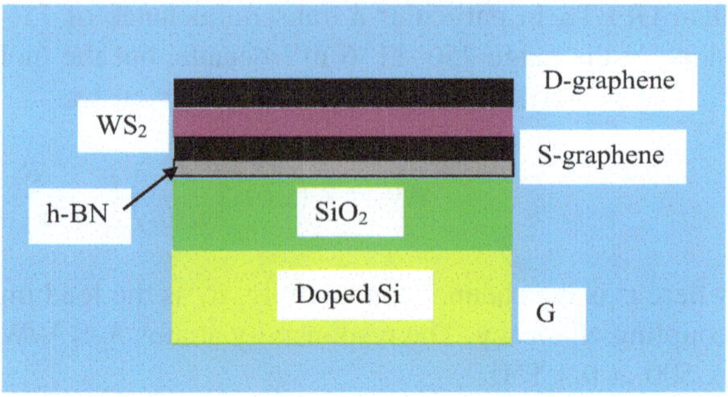

Fig. 3.23 Vertical graphene/WS$_2$/graphene heterostructure transistor

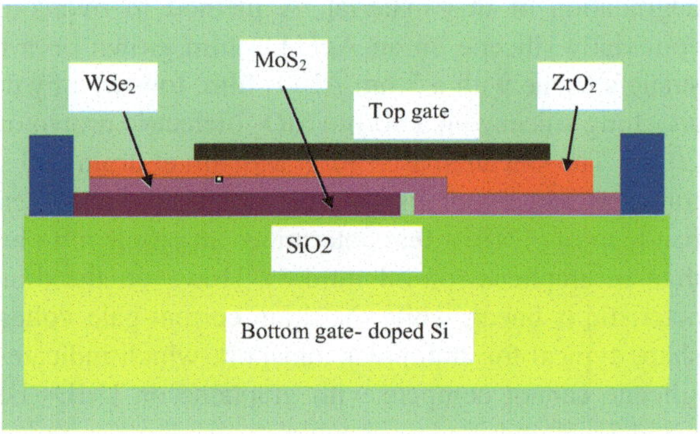

Fig. 3.24 MoS$_2$/WSe$_2$ tunable diode

Dual-gate MoS$_2$/WSe$_2$ tunnel diodes were also fabricated, the devices acting as an Esaki diode, as a backward diode, or as a forward rectifying diode, depending on biasing (Roy et al. 2015). This diode has both a top and a bottom gate, as illustrated in Fig. 3.24.

The device has selective hole and electron contacts patterned on WSe$_2$ and MoS$_2$ layers, respectively, the symmetric dual-gate configuration ensuring an effective control of band alignment via electric field distributions at the top and bottom layers. In principle, the top gate controls the WSe$_2$ layer. Because MoS$_2$ is an *n*-type semiconductor and WSe$_2$ is a *p*-type semiconductor, the structure is an *n-p* diode with a large built-in potential, acting as a backward diode if the two gates are biased at large opposite polarities. On the contrary, under a forward bias, the band offset becomes of type III, and a negative differential region (NDR) appears due to tunneling/Esaki effect. Finally, at low gate fields, the heterostructure becomes of type II staggered, the reversed tunneling current is much reduced, while a large direct diffusion current is observed. Thus, the device shows various diode behaviors at different gate regimes. The thickness of MoS$_2$ and WSe$_2$ layers are in the range 2–8 monolayers. The NDR shows a peak-valley ratio of about 2 at room temperature, with better results at low temperatures, below 100 K. The gate coupling efficiency is 80 %, but the scattering phenomena at the interface of the heterostructure can be minimized only by reducing the temperature.

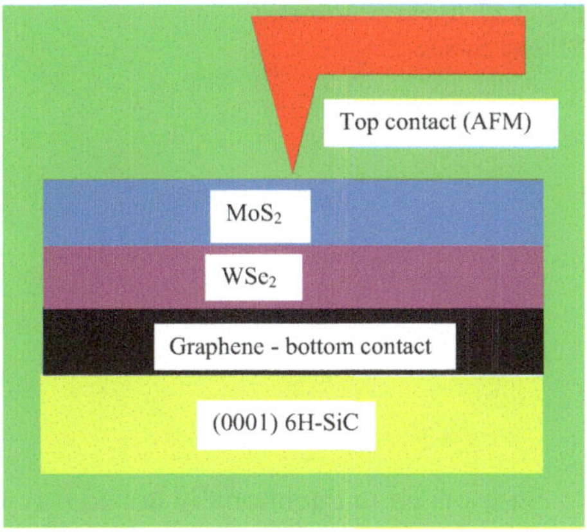

Fig. 3.25 RTD with two dissimilar TMDs

Resonant tunneling was observed in similar heterostructures, such as MoS_2-WSe_2-graphene and WSe_2-MoS_2-graphene (Lin et al. 2015). The resonant tunneling device (RTD) is fabricated by growing two dissimilar TMDs on multilayer (3 layers) graphene epitaxially grown on 6H-SiC (see Fig. 3.25).

Although the synthesis method of these RTDs is elaborated, the results are modest, structural characterizations indicating the presence of defects, wrinkles, and impurities, which prevent fabrication of high-performance devices. Note that today RTDs based on InP heterostructures work in THz with a peak-to-valley ratio > 5 and currents at mA level. On the contrary, in RTDs with two dissimilar TMDs the currents are very weak, i.e. 0.1–0.05 nA at room temperature, with a peak-to-valley ratio less than 2 for both MoS_2-WSe_2-graphene and WSe_2-MoS_2-graphene heterostructures.

A more encouraging approach is the Esaki diode configuration based on the van der Waals heterostructure formed from few-layer BP and tin selenide ($SnSe_2$), presented in Fig. 3.26 (Yan et al. 2015). The BP/$SnSe_2$ heterostructure has a broken-gap energy band offset, as can be seen in Fig. 3.27.

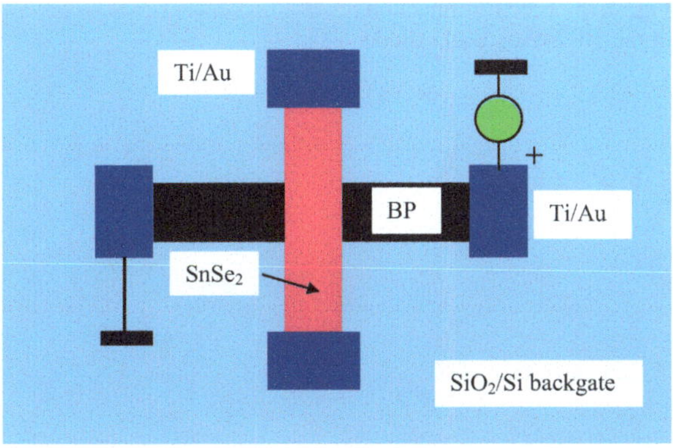

Fig. 3.26 Esaki diode formed from the BP/$SnSe_2$ heterostructure

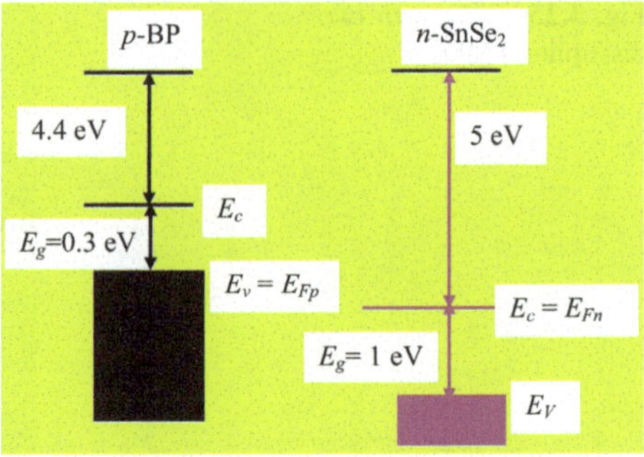

Fig. 3.27 Band structure of BP and $SnSe_2$

$SnSe_2$ is an unintentionally doped *n*-type semiconductor, while BP is a *p*-doped semiconductor. The thickness of the BP is 79 nm and that of $SnSe_2$ is 95 nm. The dry transfer method is used to assembly the two flakes. The Esaki diode, i.e. the tunneling effect through a thin insulating barrier between BP and $SnSe_2$, occurs when the applied voltage is positive because the conduction band of $SnSe_2$ overlaps the valence band of BP. A further increase of the voltage induces an injection of electrons in the bandgap of BP, reducing the current (see Fig. 3.28). The NDR phenomenon is visible at room temperature, the current reaching a peak of 150 nA at a voltage of 0.1 V. The performances of this device could not fully be appreciated, since both BP and $SnSe_2$ are unstable in air and no passivation was performed in (Yan et al. 2015).

MoS_2/graphene heterostructures can be used as nonvolatile memories with a difference factor of 10^4 between program and erase states, graphene playing the role of resistive/ohmic electrode to the MoS_2 monolayer (Bertolazzi et al. 2013). The memory is depicted in Fig. 3.29. The floating gate is isolated from the MoS_2 monolayer channel by 1 nm Al_2O_3 and 6 nm HfO_2 layers, and from the gate by 1 nm Al_2O_3 and 30 nm HfO_2. At a drain voltage of 50 mV, the drain current-gate voltage characteristic displays hysteresis, which is almost symmetric around 0 V and has a memory window of 8 V. When the top gate voltage is positive, the

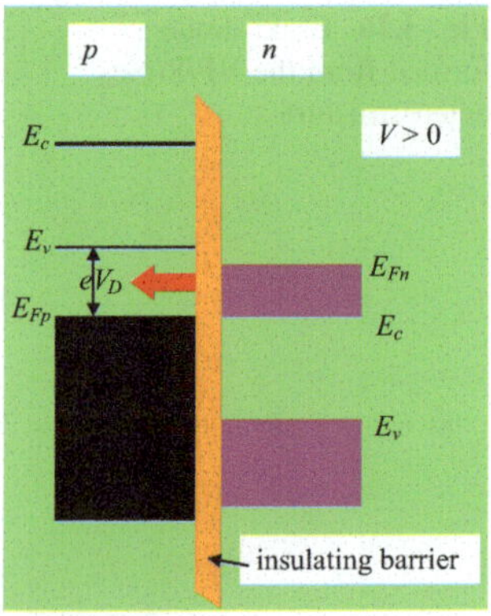

Fig. 3.28 Electron injection in the BP/$SnSe_2$ Esaki diode

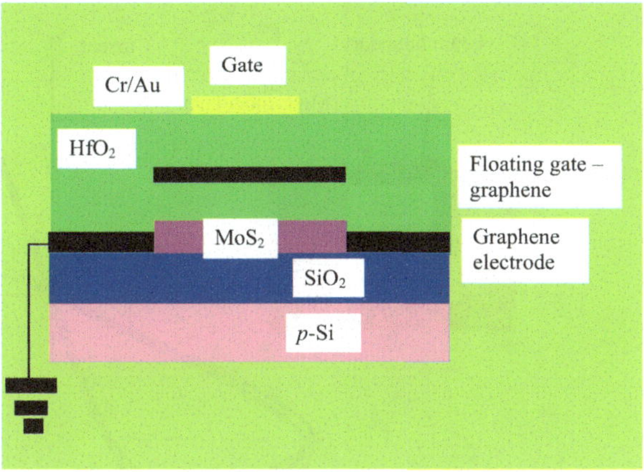

Fig. 3.29 Nonvolatile memory based on the MoS_2/graphene heterostructure

electrons tunnel from the channel to the floating gate, and remain trapped in graphene. On the contrary, when a negative top gate voltage is applied, the trapped electrons are released from the graphene floating gate, and the device is reset. This set-reset cycle is repeatable, and constitutes the nonvolatile memory effect. The two states are well separated in current by four orders of magnitude and are stable over 2000 s, the estimated charge loss being of 70 % after 10 years.

Higher memory windows, of 20 V, are obtained in a device similar to that in Fig. 3.29, except that the layers where electrons are trapped are $Al_2O_3/HfO_2/Al_2O_3$ and the back gate is used to tune the hysteresis curve (Zhang et al. 2015a). The memory device has a good retention, only 28 % charge being lost after ten years.

The memristor element was predicted long time ago, but its experimental evidence was missing until recently. A memristor is a unique circuit element, in fact the fourth fundamental circuit element, playing the same role in the circuit theory as the inductance (L), capacitance (C) and resistance (R). However, in contrast with its passive circuit counterparts, the memristor is a nonlinear circuit element, showing current-voltage dependence with a particular footprint: a pinched hysteretic behavior of the current when the voltage varies from negative to positive values (Chua 2014). The resistance of the memristor depends strongly on the history of the applied voltage. Memristors could be used in reconfigurable logic circuits or in neuromorphic systems (Kozma et al. 2012). In the latter application they mimic synapses, which are the fundamental key components of neural systems. The discovery that the memristor is the electrical analog of a brain synapse (Jo et al. 2010) has triggered attention toward neuronal networks composed of neurons (transistors) and synapses, because synapses are associated with adaptive learning.

The memristor was discovered in MIM (metal-insulator-metal) configurations (Strukov et al. 2008) where the insulator layer consists of known oxides, such as TiO_2 and HfO_2, with a thickness of less than 5 nm, and the metal layers/electrodes are Pt. There are some problems with this approach, where electrical conduction is due to oxygen vacancies that form conduction filaments, because there is no control on filament formation and on the switching voltage. In MoS_2 memristors, on the

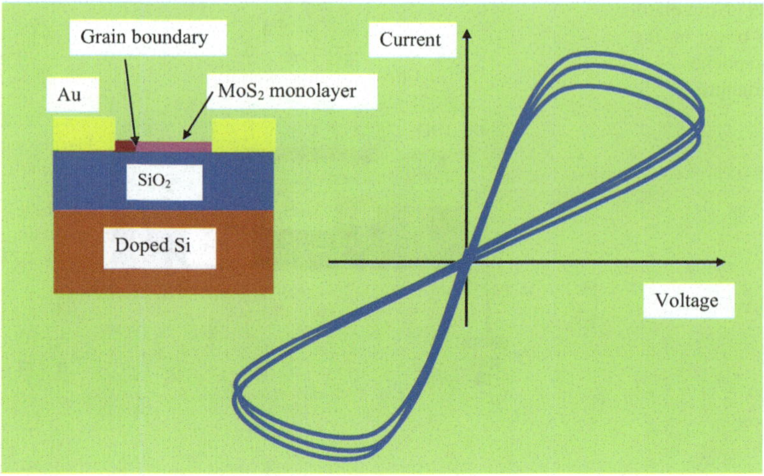

Fig. 3.30 Memristor based on monolayer MoS_2 grain boundary and a typical current-voltage characteristic

other hand, the memristive effect is caused by grain boundaries in MoS_2 monolayers, the on-off ratio taking values of 10^3 (Sangwan et al. 2015). The MoS_2 memristor is illustrated in Fig. 3.30.

The memristor, containing MoS_2 grain boundaries connected to at least one electrode, is formed by electroforming, which changes the current-voltage dependence as indicated above. Moreover, the pinched current-voltage dependence changes at each voltage sweep. The backgate can tune also the current level. The electroforming process increases the density of defects near the grain boundaries, producing mobile dopants such as sulphur vacancies. In the on state these mobile dopants migrate from the grain boundaries to carrier depleted regions, increasing the conductance and thus decreasing the resistance until it reaches the value R_{on}, while in the off state, a reverse migration of mobile dopants occurs increasing the resistance to R_{off}. In some devices negative differential resistance can be observed due to thermal rupture of the conductive filament that appears between electrodes induced by Joule heating.

The memristor model is defined by (Jo et al. 2010):

$$i(t) = v(t)/[R_{on}w(t) + R_{off}(1 - w(t))] \tag{3.14}$$

where $w(t)$ ranges between 0 and 1, corresponding to the highest and lowest conductance states, i.e. R_{off} and R_{on}, respectively. If $w(t)$ is a linear function of the flux $\phi(t) = \int v(t)dt$, then

$$i(t) = G(\phi(t))v(t) \quad (3.15)$$

Equation (3.15) is the equation of the memristor.

A MoS_2 memristor was also revealed in the 1T phase of this material (Cheng et al. 2016). In this case, the memristive action is due to the displacements of Mo and S atoms under the influence of an applied electric field, which produce lattice distortion and hence electron delocalization and conductivity enhancement. A bipolar memristor similar to that represented in Fig. 3.30 is obtained in the heterostructure $Ag/MoS_2/Ag/MoS_2/Ag$.

Electrical circuits based on 2D materials are just emerging, the large majority of circuits today being based on graphene. Inverters and a 5-stage ring resonator were already tested (Lee and Zhong 2014), but the lack of growth methods of 2D materials at the wafer scale make them less attractive for integration at the present state of knowledge. Research in integrated circuits is expected to boost as soon as the growth methods for TMDs will improved and will become widespread.

In particular, it was shown that microwave circuits could use MoS_2 as a tunable material (Dragoman et al. 2015). MoS_2 pristine flakes were deposited on a coplanar line (CPW) along which microwave fields are propagating using a drop casting method, the total thickness of the MoS_2 thin film being 3 μm. SEM images reveal the layered structure of the MoS_2 film (see Fig. 3.31). The CPW was fabricated on a 525 μm-thick high-resistivity Si wafer over which 500 nm of SiO_2 were previously thermally grown. The three metallic electrodes of the CPW were fabricated via

Fig. 3.31 SEM of MoS_2 flakes deposited on CPW (from Dragoman et al. 2015)

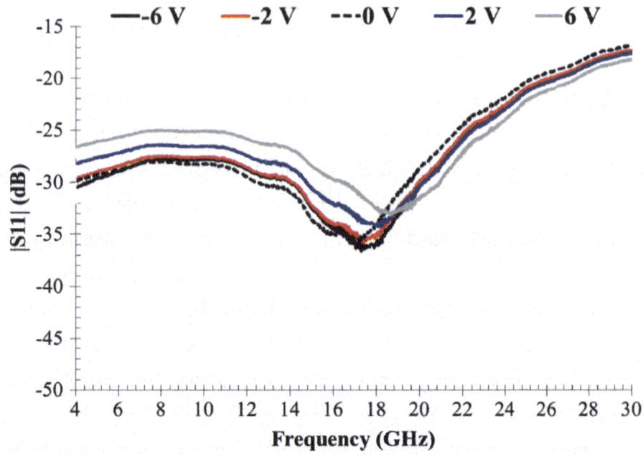

Fig. 3.32 S_{11} resonance shift in MoS_2 (from Dragoman et al. 2015)

e-beam metallization of Ti/Au with a thickness of 30 nm/300 nm, and electrochemical deposition was used to increase the thickness of electrodes to 1 μm.

Measurements of the reflection coefficient-S_{11} as a function of the applied voltage in a wide range of frequencies, which encompasses microwaves and millimeter waves, showed that the CPW has very good reflection properties, S_{11} being smaller than −20 dB in the bandwidth 4–25 GHz. If no DC voltage is applied on the CPW, S_{11} has a minimum of −35 dB at 16.4 GHz, but this resonance is shifted progressively up to 18.4 GHz when S_{11} = −33 dB, as the applied voltage increases to 6 V. The shift is due to the change of the carrier density as the DC voltage is modified. Negative DC voltages have a weaker effect on S_{11} tuning since MoS_2 is an unintentionally *n*-doped semiconductor (see Fig. 3.32). Thus, MoS_2 behaves as a tunable material/tunable resistor able to match the CPW at various frequencies.

3.3 Optoelectronic Devices Based on 2D Materials

Graphene is the most widespread and studied 2D material, its optoelectronic applications being presented in Sect. 1.4. However, for certain graphene applications in optoelectronics the lack of a bandgap is a strong detrimental factor. In deep contrast, TMD monolayers display a direct bandgap, which is very useful in optoelectronic devices, while few layers TMDs have an indirect bandgap. In the case of BP, the bandgap remains direct, irrespective of the thickness.

The optical as well as electronic properties of monolayer TMDs is determined by the electronic structure near the Fermi level, which displays two degenerate conduction and valence bands around K and K′, with spin and valley degree of freedoms. Because of the spin-orbit interaction, which removes the spin degeneracy at K and K′, the spin-up state at K becomes degenerate with the spin-down state at K′, as shown in Fig. 3.33 where states with the same energy are represented with the same color (Mak and Shan 2016). In the visible and IR spectral regions, the optical

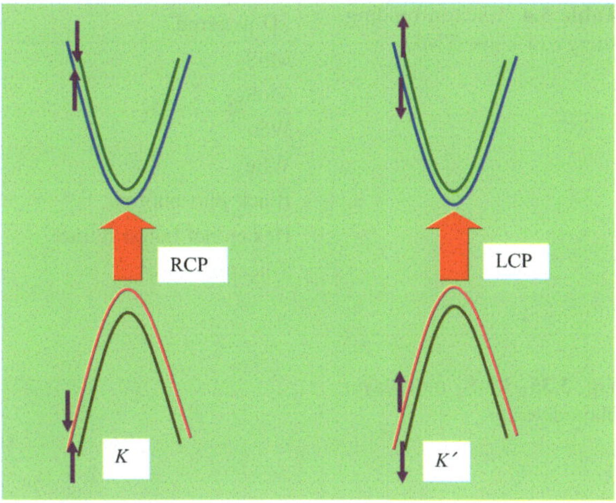

Fig. 3.33 Direct optical transitions in TMDs

properties are determined by direct optical transitions between the valence and conduction bands at K and K′. Because of the degeneracy of electronic states at the two valleys, direct optical transitions at each valley occur in the presence of either left or right circularly polarized photons, denoted in Fig. 3.33 by LCP and RCP, respectively. As such, the optical properties of monolayer TMDs depend on spins and valley states.

The absorption in TMDs in the visible and IR spectra is generally much stronger than in graphene. The low dimensionality of monolayer semiconducting TMDs generates, however, quantum confinement and dielectric screening of Coulomb interactions between charged particles, these two phenomena inducing strong excitonic effects that can be evidenced even at room temperature. The typical absorption spectrum without excitonic influence is a step-like function, the steps corresponding to photon energies for which direct transitions between different spin states at K and K′ occur, while when excitonic effects are taken into account strong

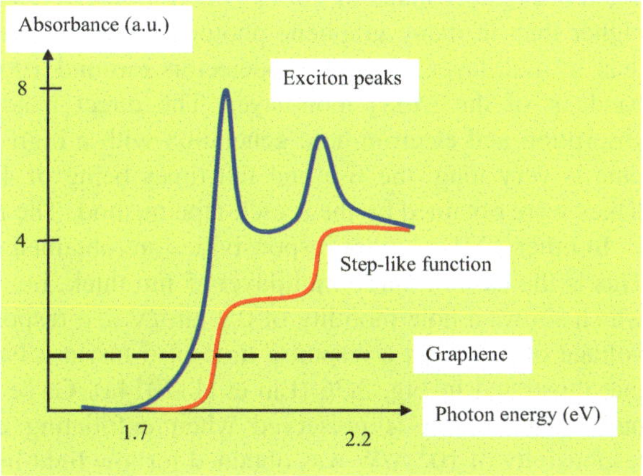

Fig. 3.34 Absorbance spectrum of MoS_2 monolayer

Table 3.4 Exciton binding energy of some TMDs

2D material	Exciton binding energy (eV)
MoS_2	0.2–0.4
$MoSe_2$	0.6
WS_2	0.4-0.7
WSe_2	0.4
Black phosphorous	0.4
Hexagonal boron nitride	0.15
Silicene	0.02

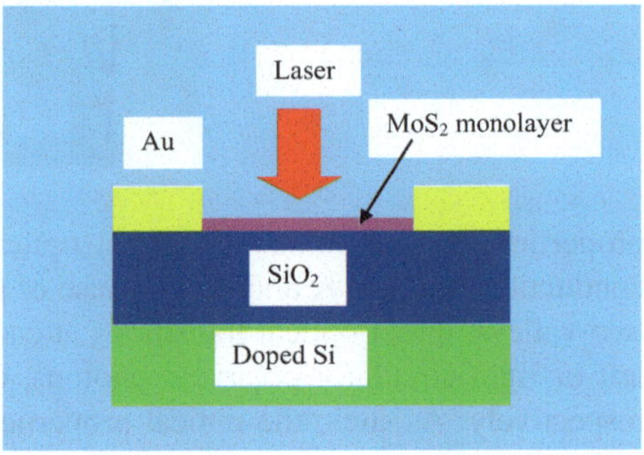

Fig. 3.35 MoS_2 monolayer photodetector

resonances occur, as illustrated in Fig. 3.34. Exciton binding energies of some TMDs are provided in Table 3.4 (Mak and Shan 2016; Castellanos-Gomez 2016).

Due to the properties mentioned above, it is expected that the photodetectors based on 2D materials overcome graphene-based photodetectors. For instance, the MoS_2 monolayer photodetector in a simple configuration of a backgate transistor excited by a laser beam (see Fig. 3.35) has shown a very high responsivity at 561 nm, of 880 A/W, and a rather good responsivity in the spectral range 400–680 nm (Lopez-Sanchez et al. 2013). The device worked at a bias voltage of 8 V and a backgate voltage of 70 V. This responsivity, which is 6 orders of magnitude higher than in many graphene photodetectors (0.5 mA/W) and 9000 times higher than in multilayer MoS_2 photodetectors (around 100 mA/W), is due to the direct bandgap of the MoS_2 monolayer. The direct bandgap is able to provide high absorption and electron-hole generation with a high yield. However, the response time is very long, the rise and fall times being of 4 and 9 s, respectively. MoS_2 flakes were obtained by the scotch-tape method. The mobility is low, of 4 cm^2/V s.

In other TMDs higher responsivities are obtained even for multilayer materials. This is the case of GaTe multilayer (5 nm thick, i.e. 10 monolayers), for example, which shows a hole mobility of 0.2 cm^2/V s, a responsivity of 10^4 A/W at a drain voltage of 5 V, and a response time of 6 ms in a backgate FET configuration, as that illustrated in Fig. 3.36 (Liu et al. 2014a). GaTe has a direct bandgap even in bulk form, which is preserved when exfoliating a single or few layers. The responsivity of 10^4 A/W was obtained for low light intensity, of 10^{-5} mW/cm^2, but

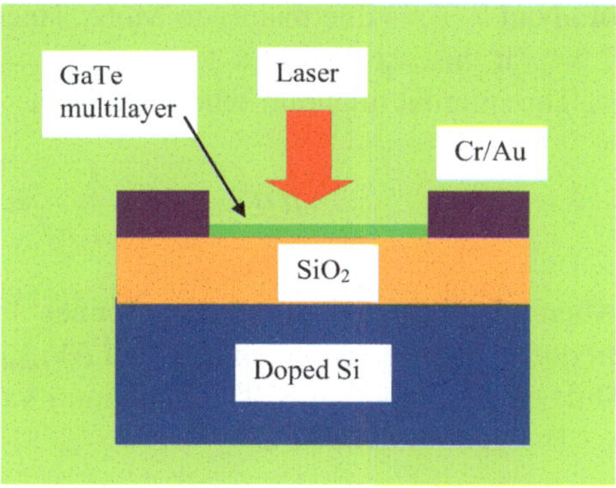

Fig. 3.36 The GaTe few layer photodetector

it decreases linearly as the light intensity increases, reaching a value of about 5 A/W for a light intensity of 10 mW/cm^2. The responsivity can be modulated by tuning the backgate voltage. The mobility of the device is low due to traps and impurities at the Si/SiO$_2$ interface, which produce significant scattering of the charged particles.

High responsivity in a backgate FET configuration with interdigitated source and drain electrodes is obtained when a graphene monolayer is transferred on a MoS$_2$ monolayer (Zhang et al. 2014); the device is schematically represented in Fig. 3.37. The responsivity at a backgate voltage of -10 V and a drain voltage of 1 V is larger than 10^7 A/W at a power density of 10^{-2} W/m^2, but decreases to about 10^4 A/W when the power density increases to 10^2 W/m^2.

The Dirac point in this device is around 10 V in dark, indicating that graphene is *p*-doped, and is shifted towards smaller, even negative values under visible light illumination. This shift can be associated to trapping of photogenerated holes in MoS$_2$, which act as an additional, positive-biased gate. The drain current behavior is similar to that of a graphene transistor, indicating that graphene, which is more conductive than MoS$_2$, dominates the carrier transport. The absorbance at 650 nm,

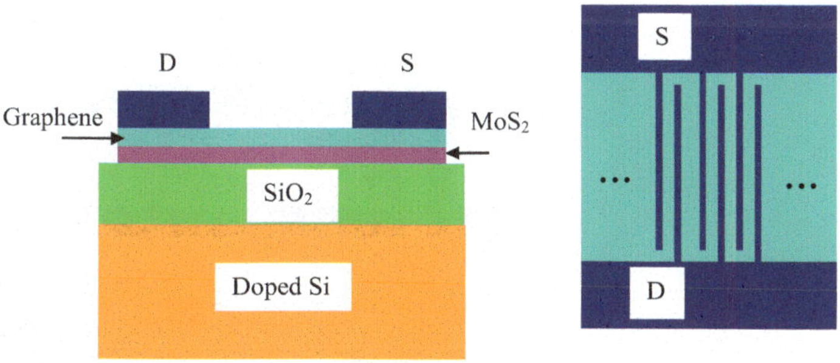

Fig. 3.37 High-responsivity graphene/MoS$_2$ photodetector

of about 7 %, is due mainly to MoS_2, since the contribution of graphene is of only 2.3 % at this wavelength.

The internal quantum efficiency of the device is given by

$$IQE = \Delta n \frac{A}{P_{opt}/hv} = \frac{C_g \Delta V_{Dirac}}{e} \frac{A}{P_{opt}/hv} \quad (3.16)$$

where A, P_{opt}, and v are the channel area, and the absorbed light power and frequency, respectively, and Δn and ΔV_{Dirac} are the density of trapped holes and the shift of the Dirac point. The maximum value of IQE is 15 % in this device.

The gain is defined as

$$G = (I_{light} - I_{dark})/(e \times \Delta n \times A) \quad (3.17)$$

and reaches 10^8 for a power density of 10^{-2} W/m^2 and a gate voltage of −10 V.

High-performance photodetectors are obtained also in vertical graphene-MoS_2-graphene and graphene-MoS_2-metal heterostructures (Yu et al. 2013). Graphene plays the role of electrodes and the heterostructures are grown over Si/SiO_2, the doped Si substrate playing the role of a backgate, which can tune the photocurrent generation. The thickness of the MoS_2 is 50 nm. These devices show a maximum internal quantum efficiency of 85 % while the maximum external quantum efficiency is 55 %, the latter parameter being defined as

$$EQE = [(I_{light} - I_{dark})/e](hv/P_{opt}) \quad (3.18)$$

The vertical graphene-MoS_2-graphene photodetector is depicted in Fig. 3.38. The photodetecting function is in this case fulfilled by the two height differences between the Schottky barriers that form at the interfaces top graphene electrode (GT)/MoS_2 and bottom graphene electrode (GB)/MoS_2. These two barriers have different heights because the Fermi energy levels in the two graphene layers differ.

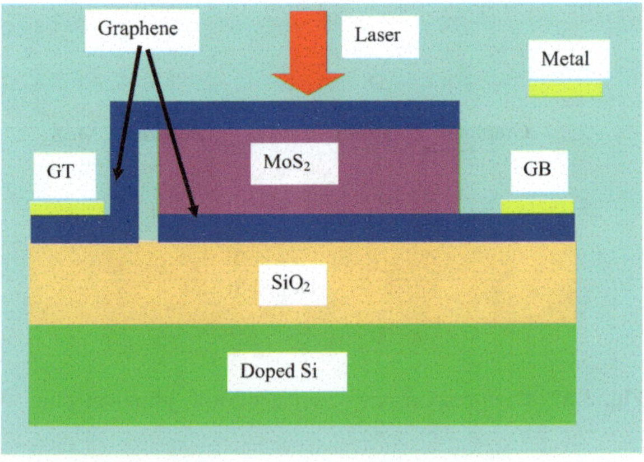

Fig. 3.38 The vertical graphene-MoS_2-graphene

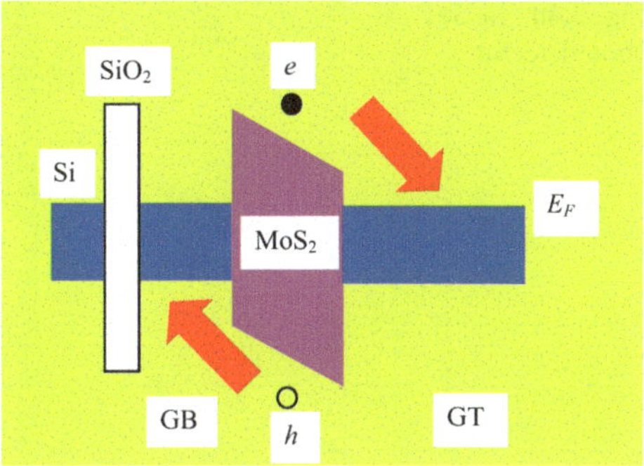

Fig. 3.39 Band diagram of the vertical graphene-MoS_2-graphene with no backgate voltage

More precisely, the barrier between GB and MoS_2 is higher than that between GT and MoS_2 due to the *p*-type doping of graphene induced by the SiO_2 oxide. Because the depletion length, of 150 nm, is larger than the thickness of MoS_2 (50 nm), the top and bottom Schottky contacts induce an overall built-in field/band slope over the vertical structure, which determines the separation and transport of photogenerated electrons and holes. As a result, the electrons are directed to the top graphene contact while the holes are directed to the bottom graphene contact (see Fig. 3.39).

The backgate modulates the carrier density in graphene, and thus the barrier height. If the backgate is negative, the barrier height of the GB-MoS_2 interface increases, the band slope increases, and the more efficient electron-hole separation leads to an increased photodetector current. On the contrary, if we apply a positive backgate the band slope decreases, reducing the efficiency of the electron-hole separation process. If the positive backgate potential is high enough, the barrier height between GB and MoS_2 becomes smaller than that between GT and MoS_2, the band slope is reversed, and the photocurrent is negative. The EQE in the vertical graphene-MoS_2-graphene photodetector varies between 15 % at a backgate voltage of −60 V and 6 % for a backgate voltage of +60 V. A similar variation of EQE is produced by changing the excitation wavelength from 458 to 633 nm.

A further engineering of the band slope can be performed by depositing an additional transparent top graphene gate on the vertical structure, isolated by 60 nm of HfO_2. In conjunction with the backgate, the top gate tunes the Dirac points of GT and GB, and thus the heights of the Schottky barriers and the band slope. In this way, the tunability of the photocurrent increases by controlling the carrier densities in the graphene electrodes and thus the screening effects. For instance, the photocurrent at backgate and topgate biases of −30 V and 0 V, respectively, is 500 times smaller than that for corresponding biases of 60 V and −6 V.

An extraordinary photoresponse is obtained using nanosheets of In_2Se_3 (Jacobs-Gedrim et al. 2014), which is an indirect semiconductor with a bandgap of 1.3 eV in the bulk form, but a direct seminconductor with $E_g = 1.82$ eV if few-layer thick. The In_2Se_3 nanosheets were transferred on an interdigitated metallic structure fabricated on 100 nm SiO_2/*p*-Si. The device, represented in Fig. 3.40, has a length of 240 nm and an active area of 0.108 μm^2. The

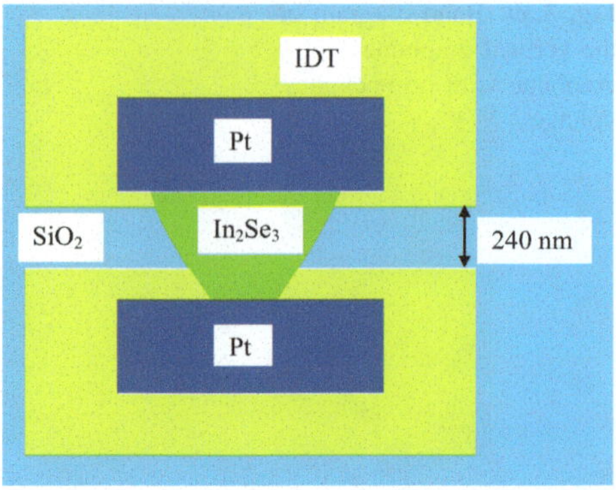

Fig. 3.40 In_2Se_3 photodetector

responsitivity at 5 V is 3.95×10^3 A/W at an excitation wavelength of 300 nm, 1.1×10^2 A/W at a wavelength of 400 nm and only 59 A/W at 500 nm. The EQE reaches a huge value, of about 160,000 % at 300 nm, the response (rise) time being 18 ms. These impressive results are due to the fact that, in addition to few-layer In_2Se_3 being a direct semiconductor and having self-terminated surfaces, which reduce recombination, in nanosheets the surface-to-volume ratio, of 2×10^8 m^{-1}, is much higher than in thin films and nanowires.

An important progress has been achieved also in coherent light emission of 2D materials. For example, WSe_2 monolayers are used for nanocavity lasers with ultralow thresholds (Wu et al. 2015). The direct bandgap WSe_2 monolayer, used as a gain medium, is transferred on a photonic crystal cavity and CW nanolasing in visible, at 739.7 nm, is obtained via optical pumping at 632 nm, with a threshold power of 27 nW at 130 K. The extreme thinnest of the WSe_2 monolayer is responsible for confining the bandgap excitons in the photonic crystal cavity. The crystal cavity is an L3 defect in a GaP membrane with a thickness of 125 nm, perforated with holes, which is transparent for the WSe_2 emission at 739.7 nm, having a quality factor of 10^4 (see Fig. 3.41).

Single photon emission was observed in exfoliated WSe_2 monolayers due to excitons situated at crystal structure defects, which act as carrier trapping centers

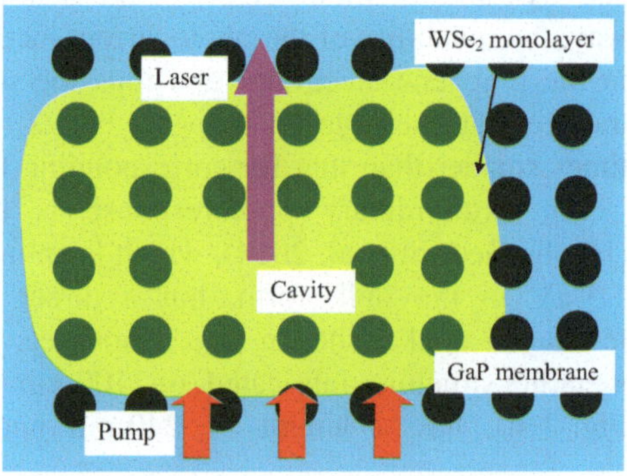

Fig. 3.41 The nanolaser based on a WSe_2 monolayer and a photonic cavity

and single photon emitters (Koperski et al. 2015). In photoluminescence spectra, the single photon emission at WSe_2 edges is identified as narrow lines at 100–130 μeV, which are clearly visible even at room temperature. These lines indicate photon antibunching, a quantum phenomena that is the imprint of single photon emitters. Photon antibunching are evidenced also (He et al. 2015) in single photon emitters based on WSe_2 monolayers grown by CVD on a Si/SiO_2 substrate. In this case, photon antibunching is demonstrated using a Brown-Twiss interferometer with a large correlation factor, greater than 0.14 and 0.2 for CW excitation and pulse excitation, respectively. The intensity of the single photon emission peaks can be controlled by applying DC electric or magnetic fields (Chakraborty et al. 2015).

Very recently, the world's thinnest optical lens was demonstrated based on few MoS_2 layers having a thickness of less than 6.3 nm (Yang et al. 2016). The few layers of MoS_2 were transferred on a Si/SiO_2 substrate and the optical path length at 535 nm was measured using phase-shifting interferometry, which allows measurements with an accuracy of 0.1 nm of the vertical optical path length. This parameter is given by

$$L_{MoS_2} = -\lambda\Delta\phi/2\pi = -\lambda(\phi_{MoS_2} - \phi_{SiO_2})/2\pi \tag{3.19}$$

where λ is the wavelength, and ϕ_{MoS_2}, ϕ_{SiO_2} are the phase shifts of few-layer MoS_2 and SiO_2 respectively. Optical path lengths in graphene monolayer samples on SiO_2 were measured in the same way. L_{MoS_2} per layer is one order of magnitude larger than $L_{graphene}$. The explanation resides in the values of the refractive indices of these 2D materials in visible: 4.4 + 0.6i for MoS_2 monolayer (Zhang et al. 2015), 2.6 + 1.3i for graphene monolayer, and 1.46 for SiO_2. As in the case of graphene, the index of refraction of MoS_2 can be tuned by applying a DC voltage.

The lens based on few layers of MoS_2 was fabricated taking into account the considerations above. The bowl-shape of the microlens, illustrated in Fig. 3.42, with a diameter of 20 μm, was fabricated using FIB. The focal length of this

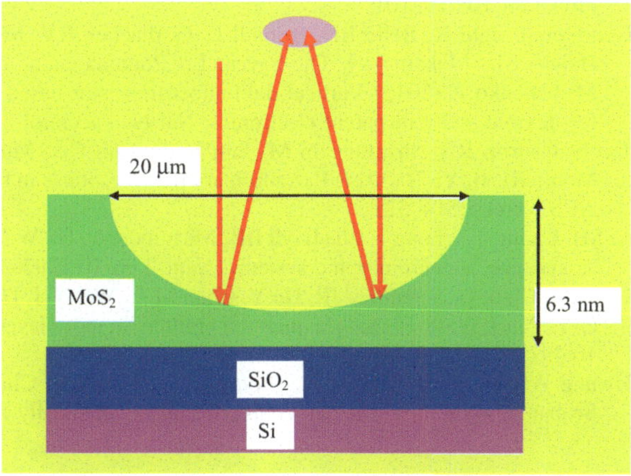

Fig. 3.42 The thinnest lens made by MoS_2

reflective concave microlens is about −250 μm. Miniaturized gratings are also reported based on the same principle.

References

Allain A, Kang J, Banerjee K, Kis A (2015) Electrical contacts to two-dimensional semiconductors. Nat Mater 14:1195–1205

Bertolazzi S, Krasnozhon D, Kis A (2013) Nonvolatile memory cells based on MoS_2/graphene heterostructures. ACS Nano 7:3246–3252

Cao W, Kang J, Sarkar D, Liu W, Banerjee K (2015) 2D semiconductor FETs–Projections and design for sub-10 nm VLSI. IEEE Electron Dev 62:3459–3469

Castellanos-Gomez A (2016) Why all the fuss about 2D semiconductors? Nat Photonics 10:202–204

Chang H-Y, Yang S, Lee J, Tao Li, Hwang W-S, Jena D, Lu N, Akinwande D (2013) High-performance, highly bendable MoS_2 transistors with high-k dielectrics for flexible low-power systems. ACS Nano 7:5446–5452

Chakraborty C, Kinnischtzke L, Goodfellow KM, Beams R, Vamivakas AN (2015) Voltage-controlled quantum light from an atomically thin semiconductor. Nat Nanotechnol 10:507–511

Cheng R, Jiang S, Chen Y, Liu Y, Weiss N, Cheng H-C, Wu H, Huang Y, Duan X (2014) Few-layer molybdenum disulfide transistors and circuits for high-speed flexible electronics. Nat Commun 5:5143

Cheng P, Sun K, Hu YH (2016) Memristive behavior and ideal memristor of 1T phase MoS_2 nanosheets. Nano Lett 16:572–576

Chua L (2014) If it's pinched it's a memristor. In: Tetzlaff R (ed) Memristors and memristive systems. Springer, pp 17–90

Das S, Chen H-Y, Penumatcha AV, Appenzeller A (2013) High performance multilayer MoS_2 transistors with scandium contacts. Nano Lett 13:100–105

Dragoman M, Cismaru A, Aldrigo M, Radoi A, Dinescu A, Dragoman D (2015) MoS_2 thin films as electrically tunable materials for microwave applications. Appl Phys Lett 107:243109

Du Y, Liu H, Deng Y, Ye PD (2014) Device perspective for black phosphorus field-effect transistors: contact resistance, ambipolar behavior, and scaling. ACS Nano 10:10035–10042

Fathipour S, Ma N, Hwang WS, Protasenko V, Vishwanath S, Xing HG, Xu H, Jena D, Appenzeller J, Seabaugh A (2014) Exfoliated multilayer $MoTe_2$ field-effect transistors. Appl Phys Lett 105:192101

Georgiou T, Jalil R, Belle BD, Britnell L, Gorbachev RV, Morozov SV, Kim Y-J, Gholinia A, Haigh SJ, Makarovsky O, Eaves L, Ponomarenko LA, Geim AK, Novoselov KS, Mishchenko A (2012) Vertical field-effect transistor based on graphene-WS_2 heterostructures for flexible and transparent electronics. Nat Nanotechnol 8:100–103

Jacobs-Gedrim RB, Shanmugam M, Jain N, Durcan CA, Murphy MT, Murray TM, Matyi RJ, Moore RL II, Yu B (2014) Extraordinary photoresponse in two-dimensional In_2Se_3 nanosheets. ACS Nano 8:514–521

Jo SH, Chang T, Ebong I, Bhadviya BB, Mazumder P, Lu W (2010) Nanoscale memristor device as synapse in neuromorphic systems. Nano Lett 10:1297–1301

He Y-M, Clark G, Schaibley JR, He Y, Chen M-C, Wei Y-J, Ding X, Zhang Q, Yao W, Xu X, Lu C-Y, Pan J-W (2015) Single quantum emitters in monolayer semiconductors. Nat Nanotechnol 10:497–502

Hwang WS, Remskar M, Yan R, Protasenko V, Tahy K, Chae SD, Zhao P, Konar A, Xing H, Seabaugh A, Jena D (2012) Transistors with chemically synthesized layered semiconductor

WS_2 exhibiting 10^5 room temperature modulation and ambipolar behavior. Appl Phys Lett 101:013107
Kang M, Rathi S, Lee I, Lim D, Wang J, Li L, Atif Khan M, Kim G-H (2015) Electrical characterization of multilayer $HfSe_2$ field-effect transistors on SiO_2 substrate. Appl Phys Lett 106:143108
Kim S, Konar A, Hwang W-S, Lee JH, Lee J, Yang J, Jung CC, Kim H, Yoo J-B, Choi J-Y, Jin YW, Lee SY, Jena D, Choi W, Kim K (2012) High-mobility and low-power thin-film transistors based on multilayer MoS_2 crystals. Nat Commun 3:1011
Koperski M, Nogajewski K, Arora A, Cherkez V, Mallet P, Veuillen J-Y, Marcus J, Kossacki P, Potemski M (2015) Single photon emitters in exfoliated WSe_2 structures. Nat Nanotechnol 10:503–506
Kozma R, Pino RE, Pazienza GE (2012) Advances in neuromorphic memristor science and applications. Springer, Dordrecht
Krasnozhon D, Lembke D, Nyffeler C, Leblebici Y, Kis A (2014) MoS_2 transistors operating at gigahertz frequencies. Nano Lett 14:5905–5911
Larentis S, Fallahazad B, Tatuc E (2012) Field-effect transistors and intrinsic mobility in ultra-thin $MoSe_2$ layers. Appl Phys Lett 101:223104
Lee S, Zhong Z (2014) Nanoelectronic circuits based on two-dimensional atomic layer crystals. Nanoscale 6:13283–13300
Li L, Yu Y, Ye GJ, Ge Q, Ou X, Wu H, Feng D, Chen XH, Zhang Y (2014) Black-phosphorous field-effect transistors. Nat Nanotechnol 9:372–377
Lin Y-C, Ghosh RK, Addou R, Lu N, Eichfeld SM, Zhu H, Li M-Y, Peng X, Kim MJ, Li L-J, Wallace RM, Datta S, Robinson JA (2015) Atomically thin resonant tunnel diodes built from synthetic van der Waals heterostructures. Nat Commun 6:7311
Liu H, Ye PD (2012) MoS_2 dual-gate MOSFET with atomic-layer-deposited Al_2O_3 as top-gate dielectric. IEEE Electron Dev Lett 33:546–548
Liu F, Shimotani H, Shang H, Kanagasekaran T, Zolyomi V, Drummond N, Fal'ko VI, Tanigaki K (2014a) High-sensitivity photodetectors based on multilayer GaTe flakes. ACS Nano 8:752–760
Liu X, Hu J, Yue C, Della Fera N, Ling Y, Mao Z, Wei J (2014b) High performance field-effect transistor based on multilayer tungsten disulfide. ACS Nano 8:10396–10402
Lopez-Sanchez O, Lembke D, Kayci M, Radenovic A, Kis A (2013) Ultrasensitive photodetectors based on monolayer MoS_2. Nat Nanotechnol 8:497–501
Mak KF, Shan J (2016) Photonics and optoelectronics of 2D semiconductor transition metal dichalcogenides. Nat Photonics 10:216–226
Miao J, Zhang S, Cai L, Scherr M, Wang C (2015) Ultrashort channel length black phosphorous field-effect transistors. ACS Nano 9:9236–9243
Mei X, Yoshida W, Lange M, Lee J, Zhou J, Liu P-H, Leong K, Zamora A, Padilla J, Sarkozy S, Lai R, Deal WR (2015) First demonstration of amplification at 1 THz using 25-nm InP high electron mobility transistor process. IEEE Electron Dev 36:327–329
Radisavljevic B, Radenovic A, Brivio J, Giacometti V, Kis A (2011) Single-layer MoS_2 transistors. Nat Nanotechnol 6:147–150
Roy T, Tosun M, Cao X, Fang H, Lien D-H, Zhao P, Chen Y-Z, Chueh Y-L, Guo J, Javey A (2015) Dual-gated MoS_2/WSe_2 van der Waals tunnel diodes and transistors. ACS Nano 9:2071–2079
Sanne A, Ghosh R, Rai A, Yogeesh MN, Shin SH, Sharma A, Jarvis K, Mathew L, Rao R, Akinwande D, Banerjee S (2015) Radio frequency transistors and circuits based on CVD MoS_2. Nano Lett 15:5039–5045
Sangwan VK, Jariwala D, Kim IS, Chen K-S, Marks TJ, Lauhon LJ, Hersam MC (2015) Gate-tunable memristive phenomena mediated by grain boundaries in single-layer MoS_2. Nat Nanotechnol 10:403–406
Schwierz F, Pezoldt J, Granzner G (2015) Two-dimensional materials and their prospects in transistor electronics. Nanoscale 7:8261–8283

Strukov DB, Snider GS, Stewart DR, Williams RS (2008) The missing memristor found. Nature 453:80–83

Sylvia SS, Park H-H, Khayer MA, Alam K, Klimeck G, Lake RK (2012) Material selection for minimizing direct tunneling in nanowires transistors. IEEE Electron Devices 59:2064–2069

Tao L, Cinquanta E, Chiappe D, Grazzianetti C, Fanciulli M, Dubey M, Molle A, Akinwande D (2015) Silicene field-effect transistors operating at room temperature. Nat Nanotechnol 10:227–231

Viti L, Hu J, Coquillat D, Politano A, Knapp W, Vitiello MS (2016) Efficient terahertz detection in black-phosphorous nano-transistors with selective and controllable plasma-waves, bolometric and thermoelectric response. Sci Reports 6:20474

Wang H, Wang X, Xia F, Wang L, Jiang H, Xia Q, Chin ML, Dubey M, Han S (2014) Black phosphorous radio-frequency transistors. Nano Lett 14:6424–6429

Wu S, Buckley S, Schaibley JR, Feng L, Yan J, Mandrus DG, Hatami F, Yao W, Vuckovic J, Majumdar A, Xu X (2015) Monolayer semiconductor nanocavity lasers with ultralow thresholds. Nature 520:69–72

Yan R, Fathipour S, Han Y, Song B, Xiao S, Li M, Ma N, Protasenko V, Muller DA, Jena D, Xing HG (2015) Esaki diodes in van der Waals heterojunctions with broken-gap energy band alignment. Nano Lett 15:5791–5798

Yang J, Wang Z, Wang F, Xu R, Tao J, Zhang S, Qin Q, Luther-Davies B, Jagadish C, Yu Z, Lu Y (2016) Atomically thin optical lenses and gratings. Light: Sci Appl 5:e16046

Yu WJ, Liu Y, Zhou H, Yin A, Li Z, Huang Y, Duan X (2013) Highly efficient gate-tunable photocurrent generation in vertical heterostructures of layered materials. Nat Nanotechnol 8:952–958

Zhang W, Chu C-P, Huang J-K, Chen C-H, Tsai M-L, Chang Y-H, Liang C-T, Chen Y-Z, Chueh Y-L, He J-H, Chou M-Y, Li L-J (2014) Ultrahigh-gain photodetectors based on atomically thin graphene-MoS_2 heterostructures. Sci Reports 4:3826

Zhang E, Wang W, Zhang C, Jin Y, Zhu G, Sun Q, Zhang DW, Zhou P, Xiu F (2015a) Tunable charge-trap memory based on few-layer MoS_2. ACS Nano 9:612–619

Zhang H, Ma Y, Wan Y, Rong X, Xie Z, Wang W, Dai L (2015b) Measuring the refractive index of highly crystalline monolayer MoS_2 with high confidence. Sci Reports 5:8440

Index

M. Dragoman and D. Dragoman, *2D Nanoelectronics*,
NanoScience and Technology, DOI 10.1007/978-3-319-48437-2

The manufacturer's authorised representative in the EU is Springer Nature Customer Service Centre GmbH, Europaplatz 3, 69115 Heidelberg, Germany. If you have any concerns regarding our products, please contact ProductSafety@springernature.com

Printed and bound by CPI Group (UK) Ltd, Croydon, CR0 4YY
30/09/2026
02225567-0010